Snakes and Ladders for Property Professionals

How to be a Smooth Operator in the Property Industry

Frances Kay
with additional material by Tanya Ross

2007

T0326030

Routledge
Taylor & Francis Group

LONDON AND NEW YORK

First published 2007 by Estates Gazette

Published 2014 by Routledge
2 Park Square, Milton Park, Abingdon, Oxon OX14 4RN
711 Third Avenue, New York, NY 10017, USA

Routledge is an imprint of the Taylor & Francis Group, an informa business

© Frances Kay and Tanya Ross, 2007

ISBN 978 0 7282 0503 1 (pbk)

Notices
Knowledge and best practice in this field are constantly changing. As new research and experience broaden our understanding, changes in research methods, professional practices, or medical treatment may become necessary.

To the fullest extent of the law, neither the Publisher nor the authors, contributors, or editors, assume any liability for any injury and/or damage to persons or property as a matter of products liability, negligence or otherwise, or from any use or operation of any methods, products, instructions, or ideas contained in the material herein.

Typeset in Palatino 10/12 by Amy Boyle, Rochester
Cover design by Rebecca Caro

Contents

Part Five — Appendix — Case Studies

Introduction

The property and construction industry is a strange beast — on the one hand, it's a huge part of the UK economy, but on the other, it can be a low-profile, low-margins business for many. It doesn't often get into the news — unless something goes catastrophically wrong, of course.

This book aims to give those who are about to — or have just — entered the industry some insight into some of the common problems. It also offers suggestions and guidance on self management, making connections, developing good relationships and managing others. In essence it helps them make the most of the opportunities that abound in this vibrant and exciting profession: avoiding the snakes and climbing the ladders.

Introduction

Part 1

Self-management for People in Property

Getting started

The man who has confidence in himself gains the confidence of others.
Hasidic saying

Whoever you are, whatever you do, as a member of the property profession, you are part of an impressive group of people. You may be newly qualified and in your first job, have recently started a new job in a more senior position or be well on your way up your chosen career ladder. Whatever your situation, you are certainly charged with achieving particular things. To achieve anything you need to be organised, and to be able to get on with other people. If you hold a supervisory or management position you will need to be in control, not only of yourself, but your staff will need to be well organised too.

Climbing the slippery pole in any profession is not easy. As any busy property professional knows, there will be plenty of hard work ahead of you and you will spend many hours in the workplace. Wouldn't it be nice if the majority of that time was enjoyable? While this cannot be guaranteed, you can take some steps towards making sure that your job is closer to a dream than a nightmare. Here are a few points to consider.

If you are starting a new job, what sort of probationary period do you have to serve? The purpose of a probationary period is to give both you and your new employer time to decide whether you are right for each other. You may be well qualified and more than able to do the job, but that now you've started you may discover that you don't really like the job or want it. This could be for any number of reasons: you may not have enough (or any) autonomy. Maybe size matters. Do you like working for a large organisation? If you do, it will help if you're good at surviving office politics. If you hate the idea of the corporate culture, maybe you would be better suited working for a smaller company where your talents will quickly be noticed. From the outset, pay attention to how you feel in the workplace and how other people react towards you. If you feel unsure, use the probationary time wisely. Give plenty of thought in those first couple of months to whether you are happy and your boss is happy with you.

Before you are able to shine at work and make the best of your opportunities, consider the likely expectations of the people you will be working with. Everyone prefers to work with people they can get on with and an ideal colleague would be someone who is:

- positive and enthusiastic
- able to see the big picture
- capable of achieving his/her own goals
- well organised and self-disciplined
- a good decision maker
- provides honest feedback
- is fair and doesn't have favourites
- open-minded and curious
- a good listener (and is available to listen)
- knows and takes an interest in colleagues
- is a good communicator
- shows confidence and gives credit
- keeps people informed
- acknowledges his/her own mistakes and weaknesses
- shares experience and helps others.

How many of these attributes do you possess? Be honest when you think about it. I hope the people you work closely with possess some of them too.

Top tip for property professionals

Make it your business to discover what is most important to your colleagues. It will be time well spent.

If you're a new recruit, hold off giving out too much information about yourself in those first few days. Best to be tactful, keep quiet and listen.

Smile and let others do the talking. Try to figure out the office hierarchy by observation — it won't be as shown on the organisational chart. Read anything and everything you can lay your hands on. Should you be joining a team, treat everyone equally unless there is a clear leader of the group. If you are given the opportunity to have extra training, it is wise to accept. Everything you learn is another string to your bow.

You may be starting a new job, but what if it is not your first? Are you being promoted by your current employer or moving on to a new one? Assuming it is the former case, you should remember that people already know you. But your position in relation to others will, and should, change. You are actually moving on while staying put — if that doesn't sound paradoxical. Because of your new role, you may have to create a bit of 'distance' between you and your former colleagues. Existing relationships and friendships shouldn't dictate the way things will work in future. You may be part of the same team, division or department, in which case you'll need to give consideration to how you act in future. Don't automatically abandon old alliances because of new circumstances — they could still be useful.

Should you be starting with a new employer, and a new organisation, your learning curve will be much steeper. Everything will be new. Discretion and caution are the best tactics here.

Top tip for property professionals

Try to match your approach to the actual circumstances and be realistic about the situation you are in.

Making an impact

When you start your job there will be lots to get you head around. A good tip is to try to get a meeting with your boss early on. Even if it is just to confirm your role and priorities, it is wise to set up an effective communication procedure between you. Does he want you to report to him regularly? If so, what is the best method? Be sure to ask appropriate questions — ideally ones which demonstrate your knowledge and intellect. The purpose of an initial meeting is to help make the first few days go smoothly for both of you.

If possible arrange to be introduced to other key people. Your work will probably involve contact with other people — possibly in another department, they may be above or on the same level as you. Some research about what these people do, and how they fit into the organisation, will demonstrate a positive approach and help to cultivate a good working relationship from the outset. When you are new, don't be afraid to ask questions. It is the one time in a job when most people will be prepared to offer you advice and information. Find out what issues are important to them and what they would most like to get out of working with you.

With a bit of preparation, you will create impact wherever you go and with everyone you meet. This will get you off to a good start. Put simply, this means:

I   instigate social activities outside work
M   manage your time effectively
P   present yourself well
A   ask questions
C   contribute ideas
T   think before your speak.

You will need to allow time to get to know your colleagues — how they work, their strengths and weaknesses. This can't be done in five minutes. While some tasks can be accomplished quickly, people related activities take a little longer. When properly managed, working relationships can be extremely rewarding. Don't be put off if some people are particularly nervous of change. They may be shy, insecure, feel threatened by newcomers, or envious of your success. Should you be working with someone who has held your position for a long time (as can happen in the case of an acquisition or merger), they will want to show a bit of control, no matter how senior (and new) you are. Observation and information-gathering are crucial in the first few weeks. Watch how people work. If it seems that one or two colleagues are being difficult, they are probably just trying to make their mark. There is every likelihood that once they get to know you they will feel less threatened and calm down.

Establishing good working relationships is essential because you may be reliant on others to deliver projects, or achieve results. Good working relationships take time, because you are trying to develop trust, earn respect and build up confidence. If you can show colleagues that you are 'on their side' they will soon become allies and possibly in time friends.

> **Top tip for property professionals**
>
> **Avoid making and acting on unwarranted assumptions about people early on. Suspend reaction in order to avoid being judgmental.**

It is a simple fact: those who are well organised give themselves a head start in everything that they do. Productivity, effectiveness, hitting targets — all important aspects of the property profession - are improved by good strategy, preparation and planning. Anything less hinders achievement, and promotes a view that you are less than efficient, led by events rather than directing your own destiny.

> **Top tip for property professionals**
>
> **If you can —** *plan the work and work the plan.* **You not only need a plan, you need to develop a method for smart working.**

S	Set task times	Divide your day/week into sections. If Mondays you want to be at your desk — avoid meetings that take you out of your office. If you like Fridays to catch up with end of week tasks — block out the time to do this.
M	Make goals	Clearly defined objectives help focus the mind and keep you motivated. Avoid setting yourself unachievable deadlines.
A	Ask for help	Never muddle through. Delegate anything you can. Enlist expertise of others whose skills complement your own.
R	Reflect:	rather than react. Avoid committing to anything until you have all the facts — a hasty decision could lead to unnecessary stress.
T	Think	— use your brain: Never be afraid to leave a task if you are stumped. Like exam techniques, if you don't dwell on it but switch to another task, by the time you return to the problem your subconscious may well have a solution.

Prioritising

The ability to prioritise is essential.

Top tip for property professionals

One senior property executive had to decide which were the most important tasks for him to tackle. Each day he made a list of things he wanted to get done. He divided his list into categories 'A' and 'B'. He tore the list in half. He put the 'B' list into the waste bin and kept the 'A' list. He then divided the 'A' list into A and B categories and repeated the process. After three attempts he arrived at the matters most urgently requiring his attention and dealt with them straight away. This may seem a bit drastic to some people (though by all means give it a try) — but it shows how to approach things in a systematic way.

The ability to prioritise is what all successful property professionals should be able to do. Successful and effective people develop the habit of doing things they *don't like*. If you can work out the important from the unimportant, you will feel more in control and work more efficiently. Important things require quality time. Urgent things have to be done quickly otherwise problems will result.

If you are used to keeping lists, and work from a 'To Do' list, assign each of your tasks into a category such as urgent, important, non-urgent but important, and finally neither urgent nor important.

- If something is both urgent and important — put it to the top of your list.
- Deal with the urgent jobs fast because they are the firefighting, crisis management things.
- Spend as much time as possible on 'important' but 'not urgent' tasks as they are the ones that have the most impact on your work or business.
- Most of the things in 'not urgent' 'not important' category are best outsourced or ignored.

To be a well organised and efficient property professional you should strive to make the most of your personal strengths and also take advantage of technology. If you are not familiar with aspects of the technology you are expected to use, take advantage of any training you are offered at the earliest possible opportunity. Make sure you behave in an appropriate and professional manner at all times — be polite, punctual and keep up to date with your work.

Top tip for property professionals

A young assistant surveyor, at an early stage in his career, was asked by his Senior Partner to attend a meeting which was being held in Conference Room C at 9.30am. When he arrived he found the door was locked. He tried to enter, but could not gain admittance. He asked several people whether the meeting had been relocated but no-one knew anything about it. Some time later he saw his boss and explained that he had been unable to attend as he could not find the meeting. He was told that the meeting had taken place, in Conference Room C at 9.30am. The young man assured the Senior Partner that he had tried to get in. His Manager advised him that the procedure in this department was to lock the door at the allotted starting time. The young surveyor was always punctual for meetings after that.

Every aspect of your work can be made more effective if you organise yourself and prepare efficiently. It may need some thought, it may even be difficult. But one thing is certain, practicing self-discipline is a good habit to develop in the early stages of your career as a property professional.

Top tip for property professionals

Being busy is not the same as being effective.

There are some people who feel that working all hours is a good way of showing that they are doing a good job. Over-doing things is not a good idea. Being over busy, and liking it or persuading yourself that there is no other way, indicates that you may be addicted to your job. This can be harmful and unproductive because it can develop into a vicious circle in which:

- you spend too long on tasks which should be finished more quickly because you enjoy doing them
- you find it impossible to say 'No' whatever the circumstances and however impossible it is to achieve the objectives or the deadline
- you reject offers of help
- you are poor at delegating and fear that by sub-contracting, the job will be less well done
- you develop a reputation for getting things done, which is what you desire, then even more work will be heaped upon you
- you have a tendency to pay little attention to training and development of your staff. This leads to individuals being de-motivated and frustrated, which can lead to an increase in staff turnover.

Before moving on to more details and other topics, these ground rules are a starting point for those property professionals who want to avoid the snakes and climb ladders. Successful integration into the workplace, where you are new to a job or the profession as a whole, requires thought and preparation.

Find out as early as possible what the 'do's' and 'don'ts' are in your work environment. If there is a staff handbook available make sure you are familiar with it. If in doubt about any aspect of it, ask the HR department for clarification. Start as you mean to go on, with a positive attitude and the intention of doing the best you can.

Being effective

Efficiency tends to deal with Things. Effectiveness tends to deal with People. We manage things. We lead people.

(Source unknown)

No property professional ever strives to be ineffective — but how is it that some are considered effective and command respect and loyalty, while others do not? Being an effective property professional is not just about being organised, getting your team to deliver projects on time and within budget — even if those are elements of your work on which you can be judged.

Top Tip for Property Professionals

An effective person is highly motivated and keeps the broader picture in mind while inspiring colleagues and employees to excel in their work.

For some property professionals this comes easily. They have natural charisma and style people admire. But, for others, however competent they are in their professional expertise, they do not easily command respect, loyalty or trust. They need to develop their own natural communication style and workplace relationships. This is a vital dynamic for anyone in the profession. Being effective is based on the ability to keep an open mind, learn from others, be ready to accept responsibilities and be accountable for actions.

If you are a young property professional, working long hours, trying to deliver a difficult project or reach performance targets, you are probably doing the best you can. But do you know how you are viewed by your colleagues, staff, partners or directors?

Benefits of feedback

Why not ask for some informal feedback from the team and departments you are working with on the project. You may be encouraged by what you hear. If the results are a bit negative — have a look at the words on the left hand column. Do any of them describe you? Perhaps if you were able to work towards being described by the words on the right, which encompass all the positive attitudes, life might become a little easier for everyone concerned.

Negative — ineffective behaviour		Positive — effective behaviour
evasive	E	encouraging
falter	F	forward thinking
frustration	F	fun
enigma	E	experienced
careless	C	confident
troublesome	T	trustworthy
insincere	I	inspiring
visionless	V	vision and values
ego	E	enthusiastic

You may be wondering why this issue is being described here. Quite simply, because it is so important. Should you not be aware of the negative impact of your, or someone else's behaviour, it is a vital lesson to learn.

Top tip for property professionals

Being able to recognise behaviour patterns in others may help you to avoid picking up those traits yourself.

Are there any colleagues with whom you've worked recently whom you would regard as ineffective? Do the words on the left hand column seem appropriate when describing their actions?

Negative — uninspiring behaviour		Positive — inspiring behaviour
ineffective	I	initiative
nervous	N	natural
suffering	S	sympathetic
pressured	P	punctual
insincere	I	impressive
rigid	R	relaxed
embarrassed	E	efficient

Consider the words connected to effective and inspiring above.

If a colleague was to assess your behaviour at work — negative and positive (left hand and right hand columns), how do you think you would rate? What do those around you, who work with and for you, think about your attitudes and approach in the work place?

When seeking some 360° feedback you will need to analyse the results:

- What insights has it given you?
- What surprises did you get (if any) good, bad, ugly and interesting?
- Where do you need to improve?

Top tips for property professionals

After receiving feedback, make notes for yourself regarding your

(1) talents and skills
(2) areas you are most appreciated in
(3) areas you need to freshen up
(4) skills you should consider learning

It's not important whether you are a first jobber, experienced property professional or the head of a department or company, everyone should try to develop skills which will help you do your job more efficiently and work well with other people. Some questions you could ask yourself are:

- How do you like to be managed?
- What skills do you need to be able to do your job?
- What support would help you?

Top tip for property professionals

To be effective, you and your team need to be focussed and working with purpose.

If you are currently involved on a project, how would you classify yourself and your team?

- High focus, low energy = disengaged
- Low focus, low energy = procrastinating
- High energy, low focus = distracted
- High energy, high focus = purpose — results.

Are you able to perform 100% in the work place? If you are, can you sustain that level of effectiveness for the duration of the project? Over and above that, are you able to manage yourself and inspire that principle in those around you? You cannot inspire others to give 100% of themselves, unless you're able to show by example that you also give 100%. This does not mean being in the office for extra long hours, working over the weekends, being too busy to take holidays and rushing around like the road runner.

Top tip for property professionals

To sustain the 'high energy, high focus' approach you will need to arrive at the best means of ensuring you and your team work effectively during working hours, while leading a balanced life outside the workplace.

Seeing the big picture

Any effective property professional needs to know how to amalgamate the successful running of a project or department while being alive to new opportunities for growth or business opportunities. In other words using the experience and knowledge of the present and the past. That means while being involved in — and possibly in overall charge of — a particular project, you should hold in mind the broader picture of cross-company objectives and future business aspirations.

Top tip for property professionals

Effective property professionals are able to network and collaborate internally and externally.

You should ensure you take the time to build relationships and forge alliances of interest with others both internally and externally. To the outsider it may seem effortless, but you know otherwise. It is because you are a professional and work actively to promote good cooperative relationships — whether at an informal meeting or on formal occasions — that the results are positive.

One of the great skills shown by effective property professionals is the ability to make decisions. The decision making process can be a lonely one. Everyone does it in their own way. A qualitative attitude survey among a group of UK companies carried out by Nicola Stevens & Associates showed that over the past three decades there are a number of styles used to make decisions.

(1)   logical
(2)   intuitive
(3)   compliant
(4)   hesitant
(5)   no thought.

How do you currently make your decisions? Are you likely to use one style more that another? Do you need to take a more balanced way of decision-making in the workplace?

Logical

You weigh up the evidence and make a decision based on the facts alone. This is a traditional analytical business process of making decisions. 'There are X, Y and Z options available to us at this time'. This is a normal decision-making style in a crisis situation or an emergency. It is however a somewhat limited way on which to base a decision.

Intuitive

This is where you make a decision on gut instinct. It is a well known fact that lasting impressions are made in first few seconds of meeting between strangers. Whether true or false, if you use this method to make your decisions, those instinctive first impressions will have a lasting influence which may in time prove unfounded.

Compliant

Do you make decisions that just 'go with the flow'? These decisions are the ones that are taken just to maintain the status quo or to keep others happy.

Try not to confine yourself to this method, it could be construed as a 'no thought' decision since there would have been little debate and it might be the weak option.

Hesitant

These decisions are the ones that just don't get taken. You will procrastinate and put off make up your mind till the last possible moment. The most experienced practitioners become so adept at this skill, it sometimes turns out that someone else makes the decision, or circumstances change so no decision needs to be made after all. If you hesitate because you simply don't know what to do, take a decision, then make it work.

No thought

This comes about because whatever the decision, the outcome is of no real interest or benefit to you or your colleagues. You simply leave the decision making process to other people on the basis that you will 'rubber stamp' whatever course of action they come up with. On the one hand, this may be an empowering approach. It can also cover up an indecisive manner, adding pressure on you in future, should someone in higher authority question why that particular action was taken.

Top tips for property professionals

To be effective remember that everyone is different and use this knowledge as a guide to how to interact with colleagues.

Summary:

- **Pay attention to positive and negative words and actions not only in yourself but also in others.**
- **Give full attention to your work in order to inspire your team.**
- **Be aware of decision-making skills and how they affect those around you.**

Time management

I am definitely going to take a course on Time Management ... just as soon as I can fit it into my schedule.

Louise E Boon

Whatever your job, most people experience problems at some stage in getting everything done in the time available. This is normal. However, for some these problems seem perpetually to exist to one degree or another. You know there are times when things conspire to prevent work going as planned, but a few confess to living in a state of near permanent chaos.

You probably have too much to do, and too little time in which to do it. Perhaps while coping with the urgent tasks you never get around to the really important things on your 'to do' list. If your desk is piled high with untidy heaps of papers, and you are constantly subjected to interruptions and have impossible deadlines foist upon you, time management isn't an optional extra — it's essential.

Top tips for property professionals

The effect of getting to grips with managing your time can be considerable and varied. It can:

- affect your efficiency, effectiveness and productivity
- influence how you are perceived by others in your organisation
- condition the pressure that goes with the job.

Time management should be seen as synonymous with self-management, which is why it is contained in this first section of the book. It demands discipline, but discipline reinforced by habit. Time management isn't just about time — it's about tasks — achieving, output and results.

Making it work

To be a successful property professional, you need to develop skills and habits which, when used correctly, work positively for you. These skills are not only necessary for doing a job successfully, they are also essential if you want to be seen as a competent, capable person. To achieve anything, as has been mentioned already, it helps to be organised. In the office, as well as at home, it's not good to be perceived as a headless chicken, led by events rather than being in control.

The techniques of time management are many and varied. Like most skills they cannot be learned by rote. You need to absorb and adapt these rules and apply them to your own unique circumstances — the tailored approach .

Top tip for property professionals

Every constructive habit you can develop will help. In other words, there's good news — it gets easier as you work at it. Good habits help ensure a well-organised approach to the way you plan and execute your work. Bad habits — like beds — are easy to get into and difficult to get out of.

Making time management work for you relies on two key factors: how you plan your time and how you implement the detail of what you do. There's really nothing worse than getting to the end of the day and feeling you haven't achieved enough. For many property professionals this is one of the more stressful things about life. Some of you may struggle quite a bit and have to make yourself plan ahead.

Top tip for property professionals

It's not the hours you put in, it's what you put into the hours.

The prerequisite to all actions is preparation and this is particularly important when tackling time management issues. Many tasks are involved including research, investigation, analysis, testing, consultation, communication, decision making. By focusing on what you want to achieve and allowing your mind to dwell on a positive outcome, you will be working towards greater efficiency.

Self-assessment

In order to improve anything, you need to be able to identify the current situation. This gives you a point against which you can measure your progress. It also enables you to identify the areas needing the greatest improvement. For instance, do you spend hours on the telephone, in unnecessary meetings, doing paperwork, socialising too much? Or are you just badly organised? You cannot begin to answer such questions without having a detailed knowledge of your own present working practices. In other words, where exactly does your time go?

Here is a quick test to see what sort of time management habits you have.

1	You begin each day by making a daily task list	Yes/No
2	You block out a chunk of time each day for dealing with essential tasks	Yes/No
3	Every quarter you review your goals	Yes/No
4	You rarely work late or at weekends	Yes/No
5	You delegate wherever possible	Yes/No
6	You usually get the desired outcome at meetings	Yes/No
7	Your telephone habits are good	Yes/No
8	You have no difficulty preventing interruptions	Yes/No
9	You have learned how to say 'No'	Yes/No
10	You finish all your tasks by the end of each day	Yes/No

If the answers to most of these questions are 'Yes' — then you are a property professional who is well aware of time management habits and skills. If they are mostly 'No's — carry on reading.

There are two simple ways to consider assessing accurately where your time goes. You could estimate it, in percentage terms, on a pie chart. Or you could use a time log. This will provide you with a much more accurate picture — because you will literally record everything you do throughout the day for seven whole days. If you can stand doing it for longer — say two weeks — you will learn even more. It only takes a few seconds to note things down, but it must be done accurately and consistently. (It is not all that different from being asked, by a dietician, to note for a week exactly what you eat.) For anyone who has ever done it, it's amazing how different the results are in practice to what you thought they would be. The surprises are usually that you spend far more time engaged in some activities than you think, while other tasks take up much less time than you realised.

What would you ideally like the time breakdown to be? If you know, this will give you a clear picture of what your time management goals really are. The objective is to give yourself something to aim towards. By keeping the time log, you will be able to tell progressively — as you take action — whether that action is working positively or not.

Top tip for property professionals

To make any real progress with time management, you need a written plan which is reviewed and updated regularly.

Ideally this should be a daily check showing accurately and completely your work plan for the immediate future. Some things are easy to include, such as progress meetings, monthly reports or budget forecasts. However some are less clear, and cannot be anticipated much in advance, if at all. Regular commitments of course should be included, together with a plan of anticipated activities for the coming month.

What you are attempting to do is to create a time management discipline, by providing information from which you make your choices:

- what you do
- what you delegate
- what you delay
- what to ignore
- in what order you tackle things.

Procrastination

It's amazing the number of people who are 'gold medal holders' in putting things off. If you have set your goals and it is written on your chart, or schedule, it makes it far more difficult to avoid doing something. One of the main causes of procrastination is disliking a task. It may be a personal aversion to a particular task or activity. If you are afraid of failure, you may try hard to avoid starting it or if you're a perfectionist, you will ensure that you have never quite finished it.

When you cannot appreciate the value or purpose of a particular task, it is unlikely that you will be enthusiastic about tackling it. It may be the result of bad briefing from a superior, or inappropriate delegation from a manager. Or the task may be so boring that you don't feel it's worth doing. With really huge tasks, it is essential that you break them into sections. This makes it easier to get started. If it's something you really hate, try doing it for just a short time. Even ten minutes tackling a nasty job is better than not touching it at all. At least you'll have broken through the pain barrier and made a first attempt.

When a job is both urgent and horrible — do it first. Getting it out of the way quickly means that it doesn't ruin the rest of the day. You can always reward yourself afterwards by having a treat. One of the ways some busy property professionals use for dealing with 'blocking' — the inability to tackle something — is by going 'public'. If you tell a friend or colleague about the issue, it will be far harder to avoid doing the job. They will keep asking you how you're getting on with it each time they see you.

Top tips for property professionals

To avoid procrastinating: here are some suggestions:

- Decide what is important.
- What do you want to achieve.
- By when do you want to achieve it.
- Do it according to your personal strengths.
- Make your plan for at least a month ahead.
- Get the list of goals written.
- Identify three important tasks and complete them.
- Enjoy the sense of achievement in marking a task 'done'.
- Check your time assessment to see how accurate it is.
- How much time have you saved by employing these habits.

With practice it is possible to become really adept at identifying leading tasks. What is the first thing that needs to be done? When that's been completed, work out what the next 'first thing' is and so on. If you take time to work out what your priorities are, have made sensible work planning decisions based on reasonable and thorough consideration of all the facts, you will be able to proceed with the task effectively.

Jobs that you hate doing are not necessarily the ones that take the longest. It's just that because you've put off doing them again and again, they loom large on your agenda. There's always the temptation to find another distraction rather than crack on with them. Everything then ends up taking longer than it should, because you're putting off facing the unpleasant task.

Here are a few simple suggestions you can put into practice straight away to help you avoid procrastinating. This could be remembered quite simply as the Four D approach.

Four D approach
Delay reaction

Stop and think for a few moments before agreeing to take on a job you really don't like, understand or want to do. Is the task you are committing yourself to something you know enough about? Have you been briefed thoroughly, and by someone who actually understands the task himself? So often problems occur at the briefing stage. If you wait until you have all the information you need to make an informed decision, the chances are you will make the right choice. Hasty decisions are often regretted.

Top tip for property professionals

The 4D approach — Delay/Diary/Delegate/Deadline

You should never agree to take something on for inadequate or the wrong reasons. This is one of the quickest ways of getting stressed. The result is that you delay tackling the job and can develop quite creative and innovative ways to avoid doing it. A task which is deferred does not go away, neither does it get easier to deal with. Delaying tactics only compounds the problem.

Diary decision

Blocking out space in your diary, or daily planner, reinforces the importance of tackling the job and helps you avoid putting it off. If you keep seeing, as you look in your personal schedule, that Friday morning is the time of the week you have earmarked for doing your fee notes (or some other admin task), you are much more likely to stick to it. The event is being reinforced in your subconscious and the seeds of a habit are being formed. With electronic diaries being so widely used it's quite likely the whole department will see your commitment. Where there is a degree of public knowledge about when you should be attending to the matter you'll probably be unable to get out of it.

Equally, if there is a big job that needs to be started, you should plan to spend an hour a day dealing with it. Rather than having to write a huge report all in one go, why not spend a little time each day for two weeks pulling it together? This is a far less painful way of getting something done. It is the same as 'chunking'. A task that is huge can best be dealt with in small sections. Some people may be familiar with the following question: 'What is the best way to eat an elephant?

Answer: 'Cut it into bite sized chunks.'

Delegate

Maybe one of the reasons you were unable to get to grips with the job you've been putting off for ages is that it really isn't one that you should have to deal with anyway. Perhaps you should stop trying to be 'superman' and think it through. Enlist the help of others to take over aspects of the task if that is appropriate. Isn't it true that a problem shared is a problem halved? Or that two brains are better than one? Possibly the job is something that could be done better by someone else entirely. In which case delegate it completely to the best skilled person. If outsourcing is a suitable solution, then consider taking that step.

Deadline

This is the final suggestion, if all of the other steps have been tried. You know what job you are taking on, because you have found out all about it before you said 'yes'. You've marked out time to deal with the matter in your diary and you've checked whether there is any part of the task that you can sub-contract or outsource to another person. Having considered all these, the task is still there, looming like a great grey cloud. The best thing to do is to set yourself a deadline. The most sensible ones are those that are realistic. If you agree to a ridiculous target date the one certain thing that will happen is that you will fail to achieve it. First you need to find out what the real deadline is and then plan your work accordingly. You will never succeed with the job if the deadline is impossible.

You may need to build in some contingency time for yourself. But whatever you do, you must start work immediately. If there is an outside chance that you are going to need help, or some more time, ask for this beforehand. It is a lot easier to overcome problems if they have been sensibly anticipated beforehand. No-one likes having to deal with a problem once it has become a crisis. To be absolutely certain, you should try to finish the task ahead of the deadline so as to give yourself enough time to go through everything thoroughly and re-check it.

To do list

A way to deal with prioritizing is to create a To do list that goes one step further. It grades the jobs in order of importance. You could use the method called 'Task Typing'. It is simply a way of categorizing the jobs you have to do by allocating each task a letter. Categorise the tasks by grading them A, B or C. 'A's are absolute musts. 'B's are chores that are less urgent but still important. The 'C's are those tasks which, should time allow, would be nice to get out of the way. There may even be a few 'D's on our list. Here's what to do:

- **Type A** tasks are important and urgent.
- **Type B** tasks are either important or urgent but not both.
- **Type C** tasks are not urgent but of low importance, routine jobs.
- **Type D** tasks are neither urgent or important, and can probably be ignored.

When it comes to scheduling tasks into the working day, you should follow these guidelines: At least one or two A list tasks should be completed each day. B list tasks are likely to take up the majority of your work. C list tasks should be fitted in around the more important work. If possible you should discard the D list tasks by either dumping or delegating them.

A typical working day will include a mixture of all types of task. It is best if you carry out a variety of tasks at different times of the day. You could hardly expect to work through all the A tasks, followed by the B tasks and so on. It is better if you can have periods of great concentration, followed by periods where you undertake less demanding jobs. Most property professionals (and just about everyone if they're honest) will admit to suffering from performance fluctuations during the working day. At certain times you are more effective and energetic than others. Provided you are aware of your own individual styles and cycles, you can plan to carry out the most intensive tasks at the optimum time of alertness. This is a great skill to acquire as it adds measurably to your personal effectiveness.

What you are actually talking about here are a few simple ways in which you can take control of your time management. These are good, practical tips which are easy to remember. They will help you to make the most effective use of your time.

Top tips for property professionals

Here are ten simple tips to create more time for yourself during your working day

1   Clear any backlog and keep it clear
2   Don't pretend that your door is always open
3   Leave things that will really look after themselves to do so
4   Actively manage the future before it happens
5   Get input and experience from people you trust
6   Have an informal chat rather than a meeting
7   When unexpected tasks are given to you, say No or delegate
8   Use down time (travel time) to think
9   Always have filler tasks available to deal with when waiting
10 Don't allow interruptions to erode your working day

Stress management

I don't believe people die from hard work. They die from stress and worry and fear — the negative emotions. Those are the killers — not hard work.
A L Williams — American Football Coach

The word stress has been around for centuries, but not in the context in which it is commonly used today. One of its original uses was the 17th century inventor, Robert Hooke, who recorded the concept in his experiments to test the tolerance of load-bearing materials. This seems appropriate considering this book is concerned with property and construction professionals. Today it is applied to humans who are also load-bearing entities. It is a popular study to measure physiologically as well as psychologically, the effects of applying too much stress.

For those of an apathetic nature perhaps you don't have quite enough stress. But when most people use the word the implication is a negative one. When you take on too much, you talk of being 'stressed out'. You might more literally describe yourself as being 'in distress' but that sounds somewhat archaic. Some people are probably more prone to stress than others, but its popularity as a modern day malaise is growing. Insurance companies have seen an increase of 88% in the number of claims made for loss of income owing to stress related issues. The simple answer is to stop worrying and start living. Stress can kill. Positive stress is okay while negative stress is not. To combat negative stress you need to develop a positive attitude. If you develop the habit of anticipating something good happening, it probably will. Stay away from those with negative dispositions, it only encourages stress.

Stress defined

Stress is a feeling that you are not in control. It carries with it an overwhelming belief that whatever you do, nothing will change. Stress is used by the brain to deal with emergencies. In moments of crisis, you experience an adrenalin rush which prepares you for 'fight' or 'flight' mode. You get a surge of energy which nature intended you to use for self-preservation.

Today's workplace constantly demands increased productivity and the provision of more and more technology to 'make it happen faster' is leading to digital depression. Increasing numbers of professionals are finding themselves on-call all hours of the day and night, with the expectation that every phone message, text message or email warrants an immediate response.According to recent press reports one in three professionals attributed increased stress to technology. The condition has been diagnosed as 'Digital Darwinism' — an anxiety caused by the belief that an evolutionary process is taking place and only the most technologically up-to-date will gain social and career success. How sad.

But why is stress so important to business? There are now more staff absences caused by stress than the common cold. Stress begins as a buzz that keeps you operating effectively. It is that boost of nervous energy that gives you your performance highs. At its best, stress is pressure. At its worst it is overload. The Health and Safety Executive (HSE)'s definition of stress is: 'Stress is the reaction people have to excessive pressures or other types of demand placed upon them.' Stress results when demands are too great or expectations are not met. The warning for business from the HSE is that stress is now the second biggest cause of work-related illness and therefore this issue needs to be urgently addressed.

Every year 6.5 million sick days are taken as a result of stress and according to the HSE 150,000 workers have taken off at least a month's sickness because of stress related illness. In order that you don't become part of that statistic it is essential that you effectively manage work-related stress, so that you:

- reduce the likelihood of sickness and absence from work
- improve performance
- have less frequent and less severe accidents
- enjoy better relationships with colleagues and clients
- experience lower staff turnover.

Top tip for property professionals

Stress can result when the demands made of you are too great or when expectations are not met. Most symptoms of individual and organisational stress are hidden below the surface (called the iceberg syndrome).

Iceberg model

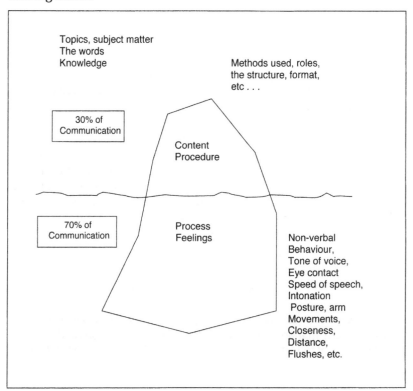

Topics, subject matter
The words
Knowledge

Methods used, roles,
the structure, format,
etc . . .

30% of
Communication

Content
Procedure

70% of
Communication

Process
Feelings

Non-verbal
Behaviour,
Tone of voice,
Eye contact
Speed of speech,
Intonation
Posture, arm
Movements,
Closeness,
Distance,
Flushes, etc.

It is only when signs such as irritability or headaches manifest themselves outwardly that other people become aware of these symptoms of stress.

When stress levels tip from positive to negative, the individual can experience some of the following symptoms:

- headache, indigestion, aching muscles
- disturbed sleep and fatigue
- change in appetite, increase in alcohol consumption or smoking
- loss of concentration, shortened temper, loss of self-esteem.

According to the *British Medical Journal,* chronic stress increases the risk of heart disease and depression. Stress can also weaken the immune system and thus your resilience to illness.

When related to an organisation, stress shows itself by:

- increased sickness absence
- significantly high staff turnover
- poor team and group working — role conflicts, interpersonal issues
- poor relationships with colleagues and clients
- back pain, RSI/upper limb disorders, bullying, harassment, poor performance.

The majority of an iceberg is hidden below the surface. This is just the same with work-related stress. Workplace stress rarely comes down to only one issue. It is a complex subject comprising many strands. The most cost effective way of reducing the risk to employee health and organisational healthiness is to find the underlying issues that are hidden below the surface. These manifest themselves through different workplace and individual symptoms.

Top tip for property professionals

The elimination of stress from an organisation is not always possible. But intelligent management can reduce risks dramatically. Usually a combination of efforts at different levels will be the most effective solution.

Work related stress is a health and safety issue of importance to everyone at work. It will have positive benefits for the company and all those working in it. Interest in this issue may begin in any part of the company but it takes the co-operation of everyone from the board downwards to put a workable and lasting strategy in place.

Some of the steps towards raising awareness among property professionals are:

- starting a dialogue, making others aware of the risks
- making positive suggestions
- identifying any department which may have particular problems
- encourage the company to carry out its own Workplace stress MOT
- explain the benefits, and the corporate responsibility for health and safety.

Top tip for property professionals

When considering individual stress, identifying the problem is essential. When you are stressed, you are on the way to losing your perspective and ability to focus. You are also running the risk of incurring health problems.

You need to take action immediately to remedy this, and to avoid getting caught in the ever decreasing circle of working harder to achieve success. Working harder is not the solution — it just won't help. Everything will take longer, you'll achieve less and your sense of stress and frustration levels will increase dramatically.

Stress symptoms

The following stress symptoms should not be ignored:

- recurring headaches
- ringing in the ears or frequent noises in the head
- regular use of self-prescribed drugs
- palpitations and chest pains
- attacks of heartburn, stomach camps, diarrhoea
- feeling that you may pass out
- getting any illness that is around
- loss of former concentration or reliable memory
- difficulty in thinking around problems
- inability to reach satisfactory decisions
- a feeling of being very low or dull

- a shut-down of all emotions except anger and irritation
- loss of sense of humour
- tears or rage appear frequently for little reason.

Defining positive and negative stress situations

Pressure at mild levels can have a positive effect. It can energise you, encourage you to perform to your best. When you reach your expected deadlines or targets or crack a difficult problem, you quite justifiably feel exhilarated and a sense of success. However, situations can quickly become negatively stressful when these pressures overpower you. You feel unable to cope and out of control. For decades, backache used to be the cause of most absenteeism from work. Stress now keeps up to a million at home on an average day — not all of whom are property professionals, I hasten to add. Up to 90% of visits to GPs are put down to stress and according to the Health and Safety Executive website *www.hse.gov.uk* over 13 million working days were lost in 2004 at a cost of £3.7 bn.

Sickness absence — both long and short term — can mask other concerns besides stress, such as poor work environment, bullying and other health issues. Should you admit (either to yourself or to others) that you have a problem, you need to identify your symptoms and make an action plan. There are a number of ways you can reduce stress levels. The most important is by learning to control your time, which is something that has already been covered in this book.

Controlling stress by common sense

When you're overworked you need to gain control. Sit down for a few moments and ask yourself what are the two or three most important things to be done today? What if nothing was done at all? What would the consequences be? Would it matter? Would it get done by someone else? One of the simplest strategies is to take control of what you can and ignore what you can't. Be happy with what you have. Sometimes it is worth remembering how much worse things could be. 'Do your best and leave the rest' is quite a sensible approach when everything seems to be getting on top of you. You're human after all. You are not designed to be perfect. When mistakes do occur, you can always learn from them.

Stress Management — Action Plans
Identifying stress

Stress can be a costly business: whether in the workplace or at home. The impact of stress on your life can be devastating both commercially and personally. In your desire to avoid snakes and climb ladders (remember that's why you're reading this book) it is helpful to be able to recognise and address stress related issues before they become unmanageable. You need to understand the causes, effects and the risks involved. Stress occurs when the pressure on any individual exceeds, or is perceived to exceed, their ability to cope. You can, interestingly, feel stressed when too little pressure is placed on you. Whether you are aware of others' stress issues, or your own, there are clear signs that will identify stress.

Top tips for property professionals

Stress shows itself in three main ways:

- psychologically   how you think
- physiologically   how you feel
- behaviourally   how you act.

The thing you should be looking out for here is 'change'.

Any change in behaviour, attitude or state of health or mind, is an indicator that someone is suffering from stress. Perhaps you, or someone with whom you are closely associated, feels excessively tired, or unwell. Perhaps you have noticed a lack of ability to concentrate or attention to detail. Perhaps there is a significant change in temperament or personal habits.

Whatever the manifestation, recognising and being aware of these signs and effects, both in yourself and those around you, is the first step towards addressing the problem. Stress can result in most serious consequences if, for example, you work for yourself, or for a small company. In these situations so much depends on each individual's personal performance.

Here are some examples of the potential effects of stress:

- valuable time lost during increasing bouts of illness or fatigue
- loss of self-motivation and belief
- lack of discipline and control which leads to
  - reduced performance and productivity
  - low self esteem
  - deadlines not being met
  - increased danger of accidents and errors.

And this is just at work. Imagine how awful you must seem at home. The knock-on effect in your personal life can be even more devastating. If you are strongly self-disciplined, particularly in the areas of time and task management, you are far less likely to suffer from the effects of stress. That is why self-discipline is so important — it gives you huge advantages over other people.

Sharing a problem with staff, associates, friends or family is one way of combating stress. Although you may not want to come across as being weak or complaining, it does make people feel valued and trusted.

In order to be able to look after others, it is essential you look after yourself. Good health is vital in keeping stress at bay. Sensible diet and regular exercise helps reduce the possibility of becoming stressed. It is also very important to be able to switch off regularly from the pressures of work.

If you find yourself losing your ability to relax, the chances are you are suffering from stress. You should be constantly alert to possible stress factors, whether in yourself or your colleagues and friends.

Top tip for property professionals

When your attitude is positive, determined, motivated and self-disciplined you will be less likely to succumb to the effects of stress.

Are you in control?

Assessing the extent to which you feel in control in your work environment is an important aid to stress reduction. Are you, for instance, able to control your flow of work? How good are you at ensuring the quality of the work you, or your staff and colleagues, produce? Do you actually pay attention to prioritizing and anticipating problems? How accommodating are you when other people come to you with stress related issues? Are you constantly aware of your own well being and that of your staff?

Consider the situations described here and see which response best fits your own style of behaviour.

You are in an important meeting with a superior. His mobile rings three times during the discussion and each time he answers it. How do you feel?

You could react furiously and throw a tantrum, storming out of the meeting. Alternatively you could sit there patiently, while he chats on and on. After all, you're being paid to wait so why worry about not getting any work done. Or you could leave discreetly, go back to your own desk and ring him up. You could then continue the meeting over the phone, or fix up a more convenient time to see him.

What would you have done? What should you have done?

The first is the reactive option. Sitting until the pressure builds up to boiling point and then erupting in fury. This would achieve very little other than to make you feel better for a moment. It would cause a bad impression with your boss and nothing positive would result.

The second choice is too passive. If you are so unconcerned about your work or your productivity, or indeed about the company, you could sit waiting forever. This doesn't give the right impression about you at all. It shows that you neither mind nor care about wasting precious time.

The third option is by far the best. It is controlled, assertive and effective. You don't lose face, and more importantly, neither does your boss. The issue here is not whether he should or shouldn't answer his phone in a meeting. What you are considering is how you react in potentially stressful situations.

A long-standing client asks you to do some work urgently. This would cause you to miss a deadline with another client. How do you deal with this?

You rush to complete both sets of work and end up delivering them way below the normal standard but maintaining the deadlines. You take the urgent work and do it as requested but let the other client down. You agree to do the work as a special favour and agree the terms — either producing less than was asked for, or setting a more realistic deadline for completion.

In this situation, what would you truthfully do? Would it be the right course of action? When dealing with potentially stressful situations it is always wise to aim for a degree of compromise.

If you grab the work and rush through it, along with your other commitments, you simply end up letting yourself down as well as your clients. There is nothing more damaging to your reputation than handing in a project that is way below the standard required.

The second option is to deal with the problem and get it done by the deadline, but in the process let down another client. This is not a wise move. The original client has every right to expect his work done on time. You appear not to have the courtesy even to contact him and explain the problem. His reaction will be one of anger or disappointment and you may lose him.

The final choice is the right one. You agree to do the work, on condition that either the client accepts a little less than he is asking, or he allows you a more realistic deadline in which to complete it.

Think about how you would be feeling in this situation. Would you feel more stressed if you selected option one? Would you worry about the effect on your other client if you took the second choice? The best option is the one where you feel happiest about the situation. The client with the urgent work feels that you care about him because you have accommodated his request, with a slight modification. There has been minimal upset and maximum gain all round. The result, a stress free working environment.

You don't have to be perfect

When faced with potentially stressful situations, it is worth remembering that you don't necessarily have to be perfect all the time. Perfectionism can bring less than perfect results and build up huge stress levels in the process.

Consider this scenario:

You are working on a time critical project. You do your best and finish work about 6pm. You are thinking about the evening you are going to have, meeting friends for drinks and dinner.

You are, again, working on an important job. You write, re-write, check endlessly and finally leave the office at 9pm in an unhappy state. You are not satisfied with the work and you've missed meeting up with friends. Result — you may have completed the task but at the expense of meeting your friends, you feel you've let them down and failed. Any advantage (warm glow) gained by doing your work well is dissipated immediately. Moral of the tale — go the extra mile by all means but not if it is going to negate any of the benefits of doing so.

What is the difference here? The first way you deal with the issue and are in control of the situation. You have done your best and that is fine. In second way of dealing with it you are trying to be perfect. What did you gain? Nothing. In an attempt to be perfect you lost out and you pile up stress for yourself in the process. The perfectionists among you feel that you must do everything yourself because no-one else's standards are high enough. You like to call the shots and exert extreme levels of quality control in most given situations.

Perfectionists feel they must know everything, handle everything and be involved with every single detail. This is far from realistic and a sure way to become stressed and lose control. Perfectionists can turn into control freaks. The result is that you often feel that you have failed because you have not reached the high standards you set yourself. In your desire to become more personally effective, surely it is better to strive to be excellent, rather than perfect? Refusing to settle for second best is a great idea, but one that you cannot always afford.

> **Top tip for property professionals**
>
> **The crucial difference between doing your best and perfectionism is being a flexible thinker. Flexible thinkers do not suffer from negative stress. They may well have some positive stress in their lives but they are able to adapt to change rapidly and create a winning solution.**

You might say to yourself, 'I'll aim to do my best, but if I don't achieve it, I won't waste huge amounts of time beating myself up for failing.' If you set yourself sky-high targets and fail to reach them, you simply end up stressed and depressed.

Studies show that perfectionists are often less productive and successful than people who don't have such unrealistically high standards. Perfectionists suffer more stress and anxiety than others. In this mode you expect everyone else around you to be perfect to. This leads to aggravation and irritation when you realise that your colleagues don't share your high ideals. Perfectionists also take far longer than others to complete tasks, either because of over-preparation or because you over-do things. Also you are frequently guilty of delaying tasks because you fear failure. All these factors contribute to building up unacceptable levels of stress.

Where perfectionists really lose out is when their 'normal' perfectionism reaches extremes. If you are driven by an intense need to avoid failure, nothing ever seems quite good enough. As a consequence you are unable to derive satisfaction from what would ordinarily be considered a job well done, or a superior performance. A constant cycle of striving, failure and self-criticism, causes stress which floods the blood with adrenaline and cortisol. These substances impair the immune system which cause you to be more vulnerable to illness — everything from flu to cancer.

The upside of perfectionism is that you do things extra well, while the downside is that you never get to enjoy a feeling of extreme satisfaction. You are either worrying about whether the job could have been done even better, or — worse — you are setting yourself even higher and more impossible standards for the next task. When attempting to deal with stress and personal effectiveness, perfectionism is a tricky pattern to break. Things should be done well, there is no argument about that. But it's the matter of the degree and motivation and how you monitor these that is the key issue here.

Perfectionism at work often shows up in over-preparation. In order to get ready for a presentation you would read and re-read your material, just to check you haven't missed out anything vital. The night before when going over the printed handouts you would find a punctuation error on the third page, dump the lot and start over again. If that doesn't build up the stress levels rapidly, what will? If you think you have to do a perfect job every time, just to avoid failure, you should consider taking some of the steps set out below.

To de-stress, have a think carefully about what 'good enough' really means in relation to performing tasks, output and productivity. Setting yourself unrealistic goals is counter-productive. In business you often hear the words 'critical success factors'. If you can identify them you should have a clear idea of what is required. Perfectionists are fantasists who imagine what a perfect job would be like and then actually aim even higher. If you are unable to get a reality check on your goals and aspirations, may be you need help from a mentor, friend or business colleague who has been in a similar situation himself. Many people suffer from 'not being able to let go' of their previously high standards. It must be understood that doing a job well — if it is slightly less than perfect — is not a failure.

Perfectionists run additional risks to that of stress and its related problems. Since you rarely allow yourself to finish tasks because they are never good enough, you are slowing yourself down. In the end you'll get left behind. The world (thanks to technology) is speeding up and the rate of change is more rapid. These days success is not dependent on delivering the perfect job. Rather it is about doing a job well enough and keeping to time and budget. You should try to make a conscious decision not to expect other people to do things perfectly. If you can develop a more flexible way of thinking and operating it will help you to combat your stress. If you ask a junior member of staff to prepare a mail shot, is it really important that each page is folded exactly in half before inserting into the envelope? Will its accurate crease really affect whether the literature is read or put in the bin? Perfectionists should strive to behave as people are supposed to do, in a fallible fashion.

Stress management — good habits and summary

Here is a checklist to discover whether you are still in pursuit of excellence and beyond. The answers should be either 'yes', 'sometimes' or 'no'.

1. When you've completed a piece of work and sent it to your boss, do you feel horribly anxious until you hear it is okay?

2. What about other people's mistakes? Say a junior member of staff makes significant errors in a piece of work, do you make a huge deal about it? Are you making a fuss because they are not as competent as you are?

3. What if you do slip up and make a mistake, do you feel an utter failure? Are you constantly requiring reassurance and praise that your work is acceptable?

4. There are a number of things you would love to try but because of your fear of not getting it right first time, you never allow yourself the opportunity to try. You deny yourself the chance of taking a risk and enjoying a challenge.

5. You are terribly organised. Your 'To do lists' cover several pages. You are unable to delegate and never ask for help, because you fear other people's standards are not as high as your own. You

over-do things and deliver work late because you are constantly trying to reach perfect standards.

What are your answers to these questions?

Mostly 'yes' — you're not thinking flexibly enough yet. You are still striving to be perfect and denying yourself pleasure and satisfaction.

If you have answered 'sometimes' to a number of these questions, you need to watch out. You're not a perfectionist but you should tackle the issues of not asking for help and never feeling satisfied with yourself.

Mostly 'no'. You're cool, flexible thinkers and are not likely to stress yourself out over unrealistic goals and standards.

Problem solving

Brady's First Law of Problem Solving: when confronted by a difficult problem you can solve it more easily by reducing it to the question: How would the Lone Ranger have handled this?

Brady

If you want to deal effectively with other property professionals, it's helpful if you can avoid complications in communicating with them. Misunderstandings can and will occur and in most cases it is because of a lack of clear communication. You will quite naturally blame the breakdown on the other person. However you may, on reflection, realise that it was something to do with the way you communicated that caused the misunderstanding.

Communication is the successful transmission of an idea from one person's mind to another person's mind. It is a process fraught with obstacles. All sorts of problems can occur here, such as:

- a lack of concentration
- a perceived prejudice about the communicator
- false assumptions about the message, or
- dislike of the communicator.

If someone wants to misinterpret whatever message they are receiving, there are plenty of opportunities to do so. It is not surprising that communication often goes wrong. One study showed that on average, people leaving an hour-long business meeting, had three to four major misconceptions about what had been agreed. If you've been in that situation when dealing with other professionals, you will have hit an internal barrier. You'll invent reasons and excuses through a process called rationalisation. You will convince yourself that the problem does not lie within you. Attitudes are a secret power working 24 hours a day, for good or bad. It is important that you know how to harness and control them. Success or failure is primarily the result of the attitude of the individual. A change of attitude can bring about an outstanding change of results.

The most successful people are deeply motivated toward their goals and objectives in life. It is rare to find successful individuals who have become successful by doing what they hate or dislike in life.

Top tip for property professionals

'Give me a man of average ability and a burning desire and I will give you a winner in return every time.'

A Carnegie

Event	Something happens
Decision	A negative outcome is decided upon. ie: 'I'm not going to get what I want.'
Actions	Subsequent actions are tinged by the decision which will affect your voice, tone and behaviour which can appear to make your decision look like the truth.
Result	The result or outcome is then a self-fulfilling prophecy. You were right.

In order to reverse the process to a positive outcome, you must observe the decisions you make. You will know you have made a negative decision when, no matter how hard you try, things still keep going wrong and you find yourself in a never ending downward spiral.

If you become adept at dealing with awkward situations, your relationships with other property professionals will be harmonious and productive. You probably experience tricky issues quite often and in many different forms. Sometimes a person has a genuine problem. Other times, it is because they don't understand some information or because of an emotional reaction that has no logic or reason behind it.

These problems, challenges or difficult situations are usually genuine for the person in their predicament. The challenge occurs when the receiver of the communication perceives it to be negative.

The purpose here is to stay calm and detached from the emotional response thereby hearing exactly what is being said. Otherwise any of the above can be perceived as an attack. As a result they become things to be feared, guarded against or, preferably, avoided altogether. The common and habitual reaction to being confronted by such situations is to defend yourself, by attacking, changing or ignoring the idea you are receiving. This makes it difficult to resolve the problem and co-operation is almost impossible. In essence two things are happening:

- there are the words of the complaint/problem
- there is the emotional charge attached to the complaint/problem.

The words will usually imply an unfulfilled need, want or expectation and need to be satisfied before the communication can proceed. The emotional charge simply has to be heard and understood.

Top tip for property professionals

It is important to remember that someone you consider a difficult person is usually delivering a group of ideas that you don't either like, agree to or know what to do with. If you remember this when you come up against a difficult situation, you will increase the chance of a mutually satisfying result.

In a situation where the automatic reaction is to defend, remember to:

(a) pay attention, listen, duplicate and understand
(b) make no assumptions
(c) listen for any free information
(d) acknowledge their ideas, repeating the essence if necessary
(e) respond to them by informing with an action or solution
(f) ask a question to ensure the situation is clear or satisfactory — if not identify the expectations before completing the meeting.

Never over explain, defend, make excuses or ignore their point of view. Empathise, don't sympathise. All they want to know is:

(a) that you fully understand their problem
(b) what you are going to do about it.

If you deal with people in this way, every situation of this kind should result in both parties being satisfied. This may look impossible. This result is not dependent on the situation being totally resolved, it is however dependent on your ability to respond and be responsible in your communication.

In summary the cycle of actions is as follows:

- keep voice pitch and tone low and even paced — match where possible
- duplicate the essence of the problem back to the other person
- tell them their options and what you can do to the best of your ability
- follow through on any actions you agree to take as a result of the communication
- if this is a situation that you found hard to handle, tell someone about it and express any residual feelings immediately
- do not make promises that you may not be able to keep.

Anger management

Let the other person express his anger. Anger is usually a short term emotion. Unless it is allowed to dissipate it will accumulate and fester. Someone who is allowed to vent their anger will be more co-operative later on. The key to this is to listen and acknowledge the emotion until it is fully released. You may get stored up anger from other things that are nothing to do with the complaint, question or criticism. Again, do not take this personally. They will probably be eternally grateful for being allowed to get a lot of other things off their chest.

Gently take back control. Once you have heard the complaint, tell them that you would like to help solve the problem and that you will do everything that you can within your responsibility, to do so. Take notes. Focus on the issue and the possible solutions, not the emotions. If the person is abusive, you should gently repeat that you want to help. Explain that you can do this better if they will tell us what they want. If appropriate, tell someone else about the situation, for instance a colleague or manager.

- Show interest by calling the person by name and letting them know that you are listening.

- Show empathy. Imagine how you would feel in the same position. Draw on your own personal experiences of times when you have been confused, misunderstood or needed an answer or explanation.

- Restate the essence of the complaint fully so there are no misunderstandings. Make sure you understand the criticism, objection, request or need.

- Consider the possibility of human error. They may have misheard something. They may not have all the facts.

- Admit the problem. If there is one, apologise.

- Ask the client/person what they want. Offer alternative solutions, not just one.

- Take responsibility for the problem until it is resolved or passed on to the correct person to sort out.

- Identify time scales. If the problem cannot be sorted out immediately, tell the client/person how long it is likely to take, even if you think this may throw up another complaint.

- Where things are difficult, it is always helpful to keep calm.

- Don't take it personally. Even if you're responsible for the error. Commit to overcoming the problem as soon as possible without resentment or blame.

- Don't give a flat 'no' answer. As much as possible offer a short explanation as to why what is being asked is not possible.

- Don't assign blame. No one cares. They just want a solution not a justification of the original error.

- Don't make promises. If you aren't sure you can deliver a promise it is wiser not make one. It will only disappoint them later on and damage the relationship even further.

- Don't lose your sense of humour, no matter what the complaint is.

Managing objections

Remember objections don't always come in the order you think that they will. Take each and every opportunity to listen and acknowledge. This is by far the best way to handle any objection.

Top tip for property professionals

Effective and positive communication begins with recognition and appreciation that each of us is unique and different. It can be achieved by tuning in your body language, tonality and words to those of another individual. Even if that individual seems to be your opposite.

All of the above is most definitely achievable but the first step towards mastery of rapport and communication is to know what you want in any communication. Once you have this established you will need three skills to work with the process.

1. Sensory acuity and awareness
   Become more alert and aware to the responses and actions of others. See more, hear more and feel more. This is a skill that can be learned.

2. Flexibility
   If you are not getting the response you want, you need to be able to change your behaviour until you get your desired outcome. You cannot expect the other person or people to change.

3. Congruence or authenticity
   This simply means that what you say and how you say it convey the same message.

Rapport or empathy is essential for establishing an atmosphere of trust, confidence and participation within which people can respond freely. Communication seems to flow when two people are in rapport, their bodies as well as their words match each other. What you say can create or destroy rapport but that accounts for only seven per cent of the communication. Body language and tonality are more important. People who are getting on well tend to mirror and match each other in posture, gesture and eye contact. Their body language is complementary.

> **Top tip for property professionals**
>
> **Successful people create rapport and in turn this helps to develop trust. This is achieved by consciously refining your natural skills. Through matching and mirroring body language and tonality you can very quickly achieve a bond with someone. This matching must be done sensitively and with respect at all times.**

Your beliefs and interests also condition what you notice. There are three main representational systems or modes through which people access, store and filter their experience. These are: *visual, auditory* and *kinesthetic*. You operate in all three modes at different times, but tend to have one preferred mode. First you should try to develop the skill of being aware of the other person's mood. Then be flexible in your own method of communicating. Visually oriented people can see what you are saying, auditory people will hear you loud and clear and the feeling (or kinesthetic) people will get to grips with your ideas.

Awareness

This is essential if you are to control a communication with other property professionals. It will enable you to:

- identify the style of communication that is preferable to the other person
- understand accurately what he/she wants — not what you think they should want
- effectively manage and respond to potentially emotional situations
- observe the impact your communication is having and alter it accordingly.

Awareness is achieved through the use of the senses and is a skill you should learn. You literally learn to see hear and feel more.

> **Top tip for property professionals**
>
> **'People like people who are like themselves.'**

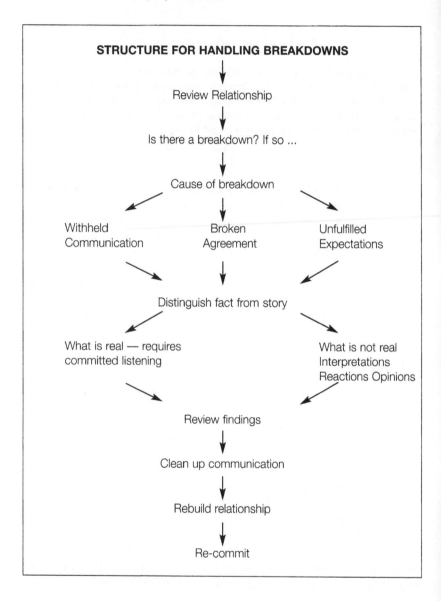

Flexibility

It is easy to talk with someone that you feel is on the same wavelength. Flexibility enables you to alter your mode of communication to suit your client. 'To speak the same language.' This may mean going outside your natural comfort zone. By using your communication skills you can defuse difficult situations and even turn them around. The Iceberg Model of Communication illustrates the need to pay attention to what is going on below the water line. Not responding appropriately and failing to deal with people's feelings is often the cause of upset. As mentioned earlier, the processes and feelings below the waterline are embedded in the past and are therefore rarely anything to do with the issue at hand. Nevertheless, unless they are dealt with they will become the cause of the breakdown.

Top tip for property professionals

The three major causes of breakdowns are withheld communications, broken agreement and unfulfilled expectations. Often it can be all three. Whichever one it is, until you can get to the bottom of it you will not be able to move the conversation on.

The diagram on p50 illustrates the route you need to take in order to resolve the issue.

Nobody likes to be the one to deliver bad news. This can often be a case of 'don't shoot the messenger'. There isn't a great deal you can do if someone is hell bent on shooting the messenger. You may feel bad about it but you must not take it personally. If you do, you will simply get into a negative spiral and end up with more upset to deal with — your own and the other person's. The guidelines here are much the same as before:

- empathise — put yourself in their shoes
- tell them the facts
- be sensitive to their feelings
- give them time for the news to sink in
- listen to them if they need to off-load
- offer your support
- end on a positive note.

This is the end of Part One — you have looked at aspects of self-management, discipline and coping strategies. Part Two deals with building relationships with other property professionals.

Part 2

Making Connections in the Property Profession

Taking the plunge

The only courage that matters is the kind that gets you from one moment to the next.
Mignon McLaughlin (*The Second Neurotic's Notebook*, 1966)

You may have been working in the property and construction industry for some time or be a relative newcomer. Whether you are experienced in networking techniques or not, it is worth considering the advantages of making new contacts with relative ease and the benefits of developing profitable business relationships. In this section of the book you will be looking at the reasons why it is necessary to be able to network. What use is it? Why should property professionals in particular need these skills and how do they go about making connections?

For a start, the profession is a gregarious one. There are countless opportunities to meet fellow professionals and socialising is something that most of them enjoy. Should you be at all shy or of an introvert tendency, you are bound to be a little fearful of what you're getting into. Networking may (in the abstract) not seem the sort of thing you feel confident about. Are you convinced that luck will be against you, that you will meet only unreasonable, unpleasant or dull people?

If you are apprehensive about the networking process you'll find it more than easy to justify not doing it. Think of the advice given about procrastination — it is equally applicable here too. You will be able to think of many valid reasons for avoiding social contact but they're not reasons really — they are actually excuses. You can't get there on time, that day you're really busy, surely you'll be able to find someone else can go, perhaps you'll do it next time — in a month, or next year. Many people faced with doing something they dislike swiftly find evasion techniques they never knew they possessed. It is best to take stock of the situation honestly and do an appraisal.

Top tip for property professionals

Networking is an essential skill. Everyone needs to be able to build business connections and there's nowhere more likely to offer great opportunities than in the property profession.

Getting started

If you really don't know how to get started on the networking front, here are some suggestions on how to do just that:

Be analytical — understand the emotional cycle

Ask yourself which situations and what occasions make you most nervous about networking. Make a list and study recurring themes and triggers. Is it, for example, that all too common fear of walking into a room full of strangers?

Self esteem inventory

Build yourself up you so that you start believing you can do it. You will manage to have a conversation with someone you have never met before. Once you have done it, it will be a positive experience.

Visualise your success

Use your imagination and picture yourself in an animated exchange with a really attractive or interesting person which results in a win-win situation for you personally or your company.

Attributes that generate success

Behave and look as though you already have the confidence you desire. Good posture and presentation help. Use positive words and phrases and speak in a confident manner when you are at your networking event.

Attitude and judgements about others

Never mind the packaging — look inside the box to see what's there. Don't allow yourself to make hasty and precipitate judgements for entirely the wrong reasons about people you've just encountered. Try to suspend reaction and take the more measured view.

Breaking new ground

Stepping out of your comfort zone is where it starts. If you do what you've always done — you'll get what you've always got. So take a deep breath and make yourself do it. You'll be glad you did.

Making that first impression count

Relax — smile — and go for it!

Top tip for property professionals

The meeting of two personalities is like the contact of two chemical substances: if there is any reaction, both are transformed.
Carl Jung (1875–1961)

If you are thinking of taking action — to start work on your personal networking journey to find some property professional connections — you will find it will enhance your confidence, self esteem and probably your career prospects too. Your goal is to make quality business contacts and keep a record of the relationships as you build them. The process starts here.

I am not really a Networking Guru. If I was, I'd be pushing you to join this club and that association and sign up for any number of amazing courses and events. You would be encouraged to rush out, shake hands with strangers and exchange contact details without any real understanding of why you're doing it. Networks matter — it's as simple as that. They enable you to access valuable resources such as solution providers and work opportunities. They also create a sense of community and rapport and allow you to share experiences with like minded people who work in your profession.

Before going much further, if you already have a list of contacts, in whatever form, review your existing network and contacts to see if they are user-friendly, effective and current. This may involve you in a lot of scrabbling around for bits of paper and old business cards stuffed in desk drawers. If this is the case, you are not alone. There are many people whose filing systems are a bit haphazard. Perhaps you are one of those people who hates clutter — and by simply switching on your computer and looking at your expertly filled data base, you will have all the names of your entire network neatly displayed at the touch of a button.

Here are some starter questions for you.

- do you communicate regularly with those in your own network?
- do you proactively seek to increase and refresh your contacts?
- do you do this formally or is it an informal arrangement?
- do you ever ask your contacts for help or offer support and advice to them?
- do you keep your network in good shape for easy access and management?

Top tip for property professionals

Networks work best according to the amount of 'give and take'. You only get out what you are prepared to put in. The best networks are information-rich, collaborative, high-trust environments. To be part of a vibrant network it is best to start simply.

To make a start, take it easy and slowly. Your network is probably already larger than you think. What about your:

- personal contacts, friends and family — people you already know
- ex colleagues and present associates from your working life
- alumni networks from the school, college or university you attended
- clients and professional contacts you meet through your job
- professional associations or social clubs you belong to.

These all contain a rich source of personalities which can yield great opportunities for you at work and beyond. You can start working on any of these groups at once. Most successful networks work on the basis of personal introductions and referrals. If you are unsure of what to do or have one or two confidence issues, the following key pointers can be used as an *aide memoire* before moving on to the next stage.

Get yourself organised with a little preparation beforehand. Preparation is essential — as in everything — personal and professional networks don't flourish or happen just by chance. First and foremost it's important to have a plan. This really matters. If you don't know why you're doing something, you won't do it well. Once you start going to networking events, you will become less worried about speaking to people you don't know at industry functions or social events you attend. To have a vibrant network you need to keep building up your range of contacts. This is the purpose of networking — to get out more.

There is no guaranteed method — or golden rule — about how to meet new people whether in a professional or purely social context. Like many things in life — it's personal. If you can try as many different methods of approach as possible you will increase your chances of success. There are lots of people who are just right for your network and how you go about meeting them is up to you. The more people you meet, the more chances there are of broadening your sphere of influence professionally. The more you grow to enjoy it, the more frequently it will lead you to like-minded people.

Top tip for property professionals

The best way to start is by just being yourself. This takes away a lot of stress and pressure about trying to make yourself appear something you're not.

Time and patience is required but straight away you will start meeting new people and enjoying new activities. With an open mind and a spirit of adventure you could expand your horizons well beyond your work place. It is best to make up your own mind about where you go, and how you make these initial approaches. Direct approaches are for those who wish to retain a degree of control. Indirect approaches to networking — even conducted via the best intentioned friends — do mean you are one step removed from the scene of operation. How can anyone but you know whether you're comfortable with a situation or not? In particular where you are trying to facilitate new contacts in your profession — it will be a bit of trial and error in the early stages. No one else can run your networking campaign better than you.

Why do you need to get out and be seen? It is the best way to make new contacts. If you've been far too task conscious up to now at work, or far too busy or too shy to make the effort, you will only build up your confidence and get some valuable experience when you get out there and just do it.

Top tip for property professionals

The best advice is to be proactive. Nothing will happen if you don't make the effort.

Wherever you begin your forays in the networking arena your first success will probably be in the most unexpected place at the least anticipated time. There are so many opportunities in the property profession to meet people, you could be out every breakfast, lunch and evening — if you could afford the time and the subscriptions to all the clubs and associations.

If you are happy pursuing casual methods of meeting new people, what about when travelling. Whether it is your daily commute, *travelling to work by bus or train*, or if your job involves international travel, *at the airport*, this could provide a great opportunity for networking. Considering the unreliable nature of many forms of public transport these days, you could well be subjected to delays. Why not use the time to strike up a conversation with someone who looks interesting while standing in a queue, or while you wait patiently at the bus stop or station platform?

Some large companies have a policy where they get involved in philanthropy or good causes. It may be that your company sponsors a charity or a fund-raising event. You could quite easily volunteer your time and energy (if you don't have much money) as a *charity volunteer*, or by participating in fundraising events. If they have the blessing of your boss, you will find this an excellent way of making new professional contacts as well as feeling good about making a contribution to a worthwhile charity project. The more involved you become in whatever voluntary work you do, the better you'll get to know others who share your interest and desire to 'make a difference'. This is where the personal touch comes in. You will be far more enthusiastic about working as a volunteer if you have passion for the cause.

Many organisations offer volunteer opportunities but some of these require a serious time commitment. Take into consideration that you will not be paid, and after all it is your professional life that comes first. Perhaps that is why the most obvious place for networking opportunities is through *work*. After all, where do you spend the majority of your time each week? Sometimes because you see the same people every day familiarity blinds you to the fact that they could be useful people to include in your network of contacts. If you are thinking of networking to enhance your career the workplace does seem a sensible place to start. You could meet a whole new crowd of people if for example you've recently moved to a new office location, or you've just been promoted to another department. Some companies have incredible opportunities for networking.

Top tip for property professionals

"Call it a Clan, call it a Network, call it a Tribe, call it a Family: whatever you call it, whoever you are, you need one.

Jane Howard, Author

Mix 'n match

If you have no fixed preference for a particular method of starting your networking campaign, it is quite okay to use any method you think you'll be comfortable with. Remember that people tend to be judgmental, you make up your mind — usually within a few minutes — whether or not you wish to pursue a conversation with someone or not. With networking, it's not a good idea to be too hasty or dismissive. Other people may be shy and not appear to advantage at the first encounter. It might be safer to keep your options open and give things time to develop. There is no limit to the number of approaches to networking you can adopt. It's your choice entirely, limited only perhaps by your resources — time and money. The most important thing is to do only what you feel comfortable with and remain in control at all times.

Given the many different types of people there are all over the world, some of you will have definite ideas as to what sort of people you like and where they may be found. There is so much opportunity and variety in the different methods and approaches to meeting new people, you will be spoiled for choice. Try to mix the methods you use, because this will optimise your chances of success, you'll build confidence faster and be able to judge which feels right for you. Make use of the opportunity to visit new and interesting places and then meeting the people along the way will be a bonus. Once you start on this process it will become easier for you to identify people with whom you have common interests, and with whom friendships may develop. Whatever you do, make sure you feel comfortable doing it.

Communicating clearly

Good communication is as stimulating as black coffee, and just as hard to sleep after.
Anne Morow Lindberg (1906–) American Author

You are about to have your first important meeting with the team on a new project. It is an essential first step in the relationship building process and your communications skills will be tested to the limit.

Face to face meetings can result in awkward pauses and initial shyness for those who are not brimming with confidence. To help you over this hurdle, you can approach the meeting fully prepared and well armed if you have a look at the following factors. In order to get your message across, think about what you are trying to achieve during the dialogue.

* What information needs to be conveyed?
* What do you want the other property professionals to do as a result?

Organise yourself beforehand. Jot down notes about your major points. Be positive and keep the message simple.

> **Top tip for property professionals**
>
> What is communication? In short, it's signalling. The transmission, by speaking, writing or gestures, of information which evokes understanding.

That's simple enough, isn't it? Straightforward in theory but in practice often fraught with danger — particularly when you have high expectations from other property professionals. Communication is not just speaking, writing or gesticulating. It's more than the transmission of information. Something else has to occur for the communication to be complete. In essence — the other parties to the communication method have to engage the brain and receive the message. When dealing with other professionals in business relationships, it's quite complex. There are plenty of opportunities for misunderstanding and miscommunication.

What happens when you open your mouth?

If you manage to insert both feet with speed and agility, you're probably nervous — don't be surprised if words come out which you seemingly have no ability to control. A conversation can go seriously wrong before you've had time to sit down.

There are some points to remember when considering the various methods of communication and some hazards to be aware of when dealing with business relationships.

- only 7% of the impact you make comes from the words you speak and that 7% comprises:
  - the type of words you use
  - the sort of sentences
  - how you phrase them.

Top tip for property professionals

If you want to make a favourable impression with the project team, consider the words, the ideas and structure of the message you wish to convey. Keep it simple if you possibly can. Always aim for clarity over ambiguity.

- commonly used words, in short direct sentences, have the greatest impact and allow the least margin for error or misinterpretation

- long words wrapped in complex sentences are confusing and best avoided: don't use jargon either — unless you are sure it will be understood by all those present

- positive statements are far more acceptable and will gain you greater advantage than negatively expressed remarks.

Effective use of your voice

Pay attention to your voice too. Tone, inflection, volume and pitch are all areas to consider. Relatively few people actually need to develop their speaking voices, but many do not understand how to use it effectively. The simplest way is to compare the voice to a piece of music — because it is the voice that is the instrument of interpretation of the spoken word.

Those who have had some training in public speaking sometimes use mnemonics as memory joggers for optimum vocal effect. One simple example is R S V P P P:

- **R**   Rhythm
- **S**   Speed
- **V**   Volume
- **P**   Pitch
- **P**   Pause
- **P**   Projection.

Rhythm

Speaking without variety of tone can anaesthetise your listener. Try raising and lowering the voice to bring vocal sound to life (and keep your audience awake). Rhythm is directly linked with speed.

Speed

Speed variation is connected to the vocal rhythm. Varying speed makes for interested listeners and helps them maintain concentration. If you're recounting a story — speed helps to add excitement to the tale. But the speed of delivery should be matched with the volume you're speaking at.

Volume

Level of volume obviously depends on where the conversation is taking place. It would be inappropriate to use loud volume when speaking in a one to one situation. However you'd probably need to increase it if you were talking in a crowded venue, such as a business reception or work area. Volume is used mainly for emphasis and to command attention — lowering your voice can add authority when telling an interesting story or giving advice.

Pitch

Pitching your voice is something public speakers do. They are trained to 'throw' their voices so they can deliver their speech clearly to their audience in whatever size or shape of the room they're speaking in. In general, it's irritating to any listener if they have to strain to hear what the speaker is saying.

In normal conversations where you need to be heard clearly (for example in restaurants where there is continual background noise as well as the hubbub of other voices) it's impossible to pitch your voice if you hardly open your mouth to let the words out. Correct use of mouth, jaw and lip muscles will produce properly accentuated words and assist with clear enunciation. Pay attention to these facial muscles otherwise your voice will be just a dull monotone.

Pause

Practice the pause. It can be the most effective use of your voice though it is often ignored. A pause should last about four seconds. It sounds like an eternity perhaps but anything shorter will go unnoticed by your listener. You can use the time to maintain good eye contact. The effect can be dynamite .

Remember the 'er' count. Filling spaces in conversation with props such as 'ers', 'ums' or 'you knows' where there should be pauses are clear signs of nervousness.

Projection

This encompasses everything about the way you come across: power, personality, weight, authority, and expertise — what some people call 'clout'. You want to build some long-lasting powerful professional connections. It pays to have some 'gravitas' in your dealings with people. Projection is an art which can be practiced. But you can learn so much from listening to experienced communicators — they have it in spades .

Reviewing your vocal skills

If possible, get a colleague or a friend to give you feedback on your voice and mannerisms. Unless you get an accurate appraisal, you could be spoiling your chances of successful exchanges. With practice you'll be surprised how quickly some of these traits can be eradicated. Once you've eliminated them and developed some of the skills suggested here, the improvement in your style of conversation and self-confidence when meeting people will be remarkable.

Top tip for property professionals

Remember, your voice is an instrument — just like your body. It is also, like your body, very flexible. You know the expression 'It's not what you say, it's the way that you say it'. That couldn't be more true.

- Be clear — use simple, easily understood words and phrases.
- Be loud (enough) so that your listener can hear you.
- Be assertive — a bright and confident tone will inject interest into anything you're saying.
- Do stop for breath — it's essential to let your listener digest what you've said — and to have the opportunity to respond.

Top tip for property professionals

Effective communication is 20% what you know and 80% how you feel about what you know.

Jim Rohn — American Businessman

Face to face encounters

The key to success is to get on to the other property professionals' wavelengths as soon as possible. By putting yourself into their shoes you'll demonstrate your ability to *empathise* with them and they'll find communicating with you easy and will show positive responses.

One of the most important aspects of communicating is to develop good *listening skills*. Lots of people are not good listeners. You are not alone if you are far more interested in what you have to say, than what the people standing next to you at the business reception are saying. Poor listening damages exchanges and that is what you are at pains to avoid. Good listening avoids misunderstandings and the errors that result from them. The behaviour of a good listener is as follows.

Attentive

A person who is listening attentively keeps a comfortable level of eye contact and has an open and relaxed but alert pose. You should face the speaker and respond to what he is saying with appropriate facial expressions, offering encouragement with a nod or a smile.

Disciplined

Adopting the behaviour of a good listener will help you establish good rapport with the other property professionals you are dealing with. It requires a degree of self-discipline and a genuine desire to take on board the message the speaker is trying to convey. You need to be able to suspend judgment and avoid contradicting or interrupting him. Postpone saying your bit until you are sure he has finished and you have understood his point.

Reflective

Reflecting and summarising — repeating back a key word or phrase the speaker has used — shows you have listened and understood. Summarising gives the speaker a chance to add to or amend your understanding. Your colleagues are far more likely to listen to you if you have let them know that you have heard what they've said by using the tactics of reflecting and summarising.

Pitfalls to avoid

• Thinking up clever counter arguments before the speaker has finished making his point.

• Don't interrupt unnecessarily or react emotionally to anything that is said.

• If the subject becomes dull or complex, don't register your disinterest by succumbing to distractions or fidgeting.

Top tip for property professionals

Listening well is as powerful a means of communication and influence as to talk well.
John Marshall (1755–1935) American Statesman

The five levels of listening skills

Level 1

The first and worst level is ignoring the speaker.

You look away, avoid eye contact and do something else altogether. (I get this sort of treatment from my family most of the time. The lights are on but nobody's there.) This is dreadful in a business context. Your colleagues will never give you the time of day again if you commit this cardinal sin.

Level 2

The second level, which is almost as bad, is to pretend to listen.

In some ways this can be quite dangerous. If you're nodding your head, and saying 'mmm, yes, aha' when you actually have no idea what's being said, you could be in for a nasty shock. Don't be surprised if you hear your colleague saying, 'So you'll run in the London Marathon next year on behalf of my favourite charity — how wonderful.' You deserved that.

Level 3

The third level listening skill is being selective.

You may well find yourself listening for key words that are of importance, such as 'successful tendering' 'phase one completion' 'on schedule'. The result is that you could miss the context of the exchange. Your colleague could have been telling you that the project is not going ahead as planned or is delayed for some reason.

Level 4

If you can develop the fourth level skill, you're doing well. This is called attentive listening.

You are focused, with positive body language, leaning forward, nodding your head appropriately and maintaining eye contact. The other members of the group know you're paying attention and this creates an atmosphere where they'll want to share valuable information and engage in serious dialogue.

Level 5

The final level is empathetic.

Empathy is the ability to put yourself in someone else's place and see things from their perspective. This takes time to achieve but it will knock the socks off anyone once you have reached it. It is the art of being able to identify mentally and emotionally with your communicator; fully comprehending the tones, pitch, body language and other subtle messages your other property professional is conveying.

It is totally exhausting to do this for any length of time but it will take your professional relationship to a much higher level rapidly. You will have included each other in the closest of possible personal networks (sometimes called a virtual team). He will consider you one of his first ports of call when information gathering or project awarding is required, and you'll willingly reciprocate.

Directing the communication cycle

Can you recall a time when you've been chatting to a work colleague, and you've looked at your watch and said 'Wow, is that the time? I must have been talking to you for ages.'

This usually happens when the two people concerned are giving each other space in their conversation. There is a feeling of ease, ideas are being passed to and fro, and a natural exchange develops. It's a bit like having a conversational game of table tennis. This is called *rapport*.

Top tip for property professionals

If you can begin the rapport building process with your property professional colleagues, you will begin to cultivate the relationship you want and your exchanges will become frequent and more valuable. You're attempting to establish the balance of listening and talking.

There are times when you'll want to find out more information. It's easy to ask too many questions and fall into a sort of 'Spanish Inquisition' situation. Conversely when responding to a question you can give away too much information. If you're on the receiving end of this from your colleagues on the project team, the relationship may not make much progress. No one likes to feel they are being 'pumped' for information. It's infuriating and insulting and you'll want to distance yourself as quickly as possible.

Only one person at a time can truly direct a conversation. One leads and the other tends to follow. This doesn't mean there is no give and take. Neither does it mean that the other party is subservient. But one of the parties should lead and there is merit in you being the one that does so, if your objective is to build a proactive professional relationship with rewards for both sides.

Opening rituals

At the start of a meeting, there are usually some general opening remarks. This sort of ritual is customary and should take no longer than a few minutes at the outset of proceedings. Watch for the moment when the chattiness should cease because if you have no real plan, the other members of the team may lead you off into uncharted waters and you find yourself heading in the wrong direction.

Someone usually starts off by saying, 'Right, shall we move on? Can you tell me ...' That person could be you. If no-one else seizes the opportunity to take control at this point you should, or else you may have lost the initiative for the rest of the meeting.

You might consider going into the meeting with a short agenda. If this isn't written down, at least it should be in your head. It could be little more than a few helpful suggestions. Perhaps you've already aired the topics for discussion in a telephone call beforehand. There is no rule here, but whatever has been agreed it does at least mean that the exchange proceeds along some agreed lines.

It also provides an element of control during the dialogue, if the conversation meanders into other areas. You could refer back to your brief by saying something like, 'we were going to discuss X next ...' and then move on smoothly to the next stage. The early part of any meeting is a key stage for your confidence; you'll feel and operate better if you get off to a planned start and you'll be able to maintain better control and direct the rest of the exchange.

Good conversational techniques

To develop a balanced style of communication, try to begin the conversation by introducing yourself and giving some personal information.

This is called the *inform* stage.

Once you've given some information, ask a direct question of the other members of your project team.

This is called the *invite* stage.

Then *wait* for his response.

On receiving this, *listen* to every word.

Then *acknowledge* and, if necessary, repeat the essence of their response.

If you achieve this cycle of communication you can repeat it many times over during the encounter to establish a good rapport between you and the other parties. It should make the time pass effortlessly and harmoniously and make your exchanges a pleasant experience. Building good relationships with other members of the project team has similarities to the dating process. You are attempting to get closer to your team members by developing the art of good conversation, so pay attention to the importance of *eye contact*.

Appropriate eye contact at all times in the exchange is essential. If while you are talking you notice that the rest of the group are looking at you with an interested expression, nodding occasionally and smiling at the right times with an alert and open posture, you're holding their attention and doing alright.

Things to look out for

However, should one of your colleagues appear to be *falling asleep* during one of your conversational gambits, it could mean that

- he's had a late night
- he's had an early start
- he's suffering from jet lag
- the atmosphere in the room is too stuffy
- or your dialogue is rather boring.

Don't wait until his head falls forward and hits the desk. If you fail to notice until you hear the crash, you're definitely talking too much.

Keep an eye out too for fidgeting, this could indicate that

- you've lost his attention
- he wants a break
- he's irritated by something you've said
- or he finds the conversation irrelevant.

Whatever the reason, it's time to shut up. Close mouth without delay and smile. Hopefully with a bit of silence you can retrieve a relationship that may have got off to a rather inauspicious start.

Should one of your colleagues start *shaking his head*, this could mean

- he wants to say something
- he doesn't agree with you
- or he simply hasn't a clue what you're waffling on about.

Again, as above, time to bring your remarks to a swift close.

If you think you've *lost his attention* completely, and he turned off, try to regain it by asking him a pertinent question. Re-establish eye contact and vary the volume or expression in your voice.

Other forms of communication

Telephone calls

These can be difficult to deal with and can often cause trouble between parties who do not know each other all that well.

First, because you can't see each other face to face, you have to rely on tone of voice. This can be deceptive. He may sound disinterested because he's talking in a low voice. It may be something as simple as the fact that he's got a sore throat, or he's trying to avoid the rest of the office hearing his conversation.

It's essential to pay attention when a member of your new project team calls. If he's on a mobile, you may well get a distortion, due to background noise, traffic, airport announcements or similar. If possible take the phone call in a private place so as to avoid even more noise coming from your end of the phone.

Voicemail messages

There's an art to leaving successful voicemail messages. It's simply this: be clear and be concise.

Don't speak too fast. If you are leaving your telephone number, slow down. Speak slowly while recording the information.

If the message you leave is either gabbled or garbled, it will be impossible for any one to return your call. It helps to leave a date and time when you record your message, so that your other property professional can respond quickly if time is critical.

Text messages

This is the perfect form of communication for quick exchanges of information.

One word of warning — don't use confusing abbreviations. If you received the following message — CU 7.30. Does that mean 'See you at 7.30pm' or 'Curtain up at 7.30pm'? Check if you are in doubt.

Written communication

The main point about written communication is that whatever form it takes, the recipient cannot see you or hear you.

Your colleague has no option but to accept what they read. You should pay particular attention to wording and expressions because if it is at all ambiguous, it is liable to be misinterpreted.

Letters

When handwriting letters put yourself in the position of your recipient.

Write neatly and clearly and make sure your spelling is correct. It helps to use a decent pen and good quality paper. Impressions count — remember.

With a personal thank you note, use the business address because it is after all a business relationship, even though you are thanking him for inviting you to a social occasion. Keep your message simple and make it easy to read. Layout is important. Avoid innuendo, sarcasm and *doubles entendres*.

Email

Much has been written about email etiquette, because this is such a popular and efficient form of communication.

You wish to email to your project team colleagues, check which address is the most appropriate. There may be confidentiality issues — particularly if your exchanges may include something other than the job in hand — and a personal email address may be more appropriate.

If he says it's okay to email to his business address, do be circumspect. Emails may not always reach the recipient directly. Some people have staff who scan emails before forwarding on to the main addressee.

Consider the likelihood that your email is going to be read by someone else, so be extra careful. Any references to someone's personal habits (his, yours or other peoples) can be seized upon and rapidly transmitted around the world by an enthusiastic prankster.

On a more practical level, email is not the medium for rambling on and on about the project, or any other subject that is dear to your heart. Keep email communication clear and short. It's no substitute for face to face contact, but it does allow for a fast exchange of information, particularly when confirming meetings or referring to matters just discussed.

Top tip for property professionals

The most important thing in communication is to hear what isn't being said.
            Peter F Drucker (1909–) American Management Consultant

---

Communication skills awareness checklist

Presence

Pay attention to the way your voice and body language are used in conjunction with the words you speak. You can convey the right impression if they are used correctly.

Relating

Don't underestimate the importance of developing your rapport building skills to get on the same wavelength of your new professional colleagues.

Questioning

When engaged in conversation with your other property professional, make sure you match your question to the situation or subject. Beware asking irrelevant questions — this will show that you've not paid attention to what he said.

Listening

Listen to everything he says attentively. Try to reach at least Level Four. If he's likely to become a significant influence in your business development strategy you should aim for Level Five eventually.

Checking

The art of glancing at your colleagues to see that they are still on your wavelength while you're engaged in dialogue. Watch for gestures and see whether they do the same when they are talking.

Manners and mannerisms

The hardest job kids face today is learning good manners without seeing any.
Fred Astaire (1899–1987) Dancer, Singer, Actor

Politeness among property professionals is desirable and often quite rare. Being courteous to colleagues, co-workers, staff and clients will significantly influence relationships in the workplace and foster a harmonious culture. Hopefully you are not a troublesome character, but go out of your way to deal pleasantly with others. It is so important to preserve other people's dignity and respect. That alone will reduce the number of difficult situations and personality clashes which people encounter time and time again. The professional colleague who rides roughshod over other people's feelings in the project team does often end up with a load of headaches — mainly caused by his own actions.

Non-verbal communication

If you are familiar with the practice of non-verbal communication it can be used effectively to *soften* the hard-line position of others:

S   smile
O   open posture
F   forward looking
T   touch
E   eye contact
N   nod

If you approach someone with a smiling face, it encourages a similar response from the other person. Just as important is your presentation and body language. The posture should be open, head upright and you should stand straight but with hands relaxed by your sides. Make appropriate gestures to show that you are welcoming the exchange

Eye contact should at all times be honest and open. Avoid staring but maintain a steady gaze when speaking to colleagues. At the same time, encouragement by nodding your head shows consensus and indicates that you are taking note of the points being made.

Polite behaviour should ensure smooth interaction between colleagues working on a team or project. If you practice giving the right signals, with luck, others will mirror your actions. After all, imitation is the sincerest form of flattery. If for some reason the project team is full of people who are backstabbing and politicking, harassment, bullying, and discrimination could follow. Sometimes a working atmosphere worsens because a new arrival joins the team. Some people do seem to specialise as virus spreaders — the ones whose arrival begins with a chill factor and then spreads like an epidemic, whether it be 'flu or resignation.

Maybe someone's approach needs a makeover? It is easier to adapt an individual's manner when dealing with others, than it is to alter an endemic organisational culture. But if the problem is not too great to be rectified, it may seem a bit simplistic — but a little praise goes a long way towards helping with these issues.

Top tip for property professionals

As Catherine the Great of Russia once said 'I praise loudly. I blame softly.'

Maintaining good morale

Good traditional praise does not go amiss in the workplace. One of the keys to retaining goodwill among professional colleagues is to foster a sense of camaraderie amongst them. The more control project team colleagues have over how, when and where their work is done, the happier they will be. Their performance improves, along with their morale. There is less confrontational behaviour — in short, everyone gains. Regardless of how advanced technology has become and what the latest equipment and gadgets enable us to do, some things are in danger of becoming too impersonal and remote.

Top tip for property professionals

There is nothing more encouraging than good manners and personal attention. Where a plague of bad behaviour, or ill manners, pervades the workplace, there is bound to be an increase in problems among professional colleagues.

Kindness should not be underestimated. If someone is showing signs of anxiety, stress or depression, they are probably feeling inadequate and undervalued. Left unchecked, this situation could spiral towards absenteeism. Work-related anxiety has knock-on effects because it doesn't go away if you ignore it. There is a marked difference between stress and anxiety — the former is caused by over stimulation and overload. The latter is usually because the person feels a failure, a poor performer or inadequate. Should there be any signs of bullying, harassment, sexual or racial prejudice towards this person, these could result in a number of physical symptoms — shortness of breath, nervous behaviour, back or headaches, eating disorders and insomnia.

Although dealing with anxiety is best done by seeking professional help, starting with a little kindness can go a long way to combat the effects. Helping colleagues cope with challenging situations by praising them can alleviate the anxiety symptoms. However severe the anxiety displayed is, there is no doubt that being kind to others will help to redress the balance. One important tip is to note how people breathe. Anyone showing nervousness, anxiety or stress, will hold their shoulders in a rigid way and have shallow breathing. Think about any relaxation techniques you have been taught, they all encourage deep breathing, shoulders down and arms held loosely at your sides. If you are able to observe your colleagues breathing and posture, it should be quite evident who is not relaxed and at ease. Start your kindness campaign with them.

If you are expected to deliver bad news — some criticism to someone who is under-performing — it is still possible to do so without hurting someone's feelings. Sweeten your message by making it clear that you are trying to help them and work out what it is you want them to do differently. Have the conversation with them at an appropriate time and place, so as to avoid any embarrassment to them in public, particularly if it is a sensitive issue.

You need to be sure that the person knows what you are talking about — whether it is their performance over the past few months or a piece of work they have delivered to you that morning. Be specific and keep it concise: 'your poor timekeeping is causing a problem for the department' is precise. Reminding them that they have been a good role model in the past will reinforce their identity and encourage them to think positively about themselves.

Top tip for property professionals

If you are able to dispense honey rather than vinegar, you might influence your colleagues towards a change of mood. Smiling is the first step, and being happy around them should help to encourage a warmer atmosphere.

Professional colleagues whose motivation levels are high are not likely to be suffering lack of self-esteem or stress. Should grumpiness pervade the office, pay careful attention. Beware the chill factor — should moody people be allowed to simmer, you will find this contaminates the team fairly swiftly. The benefits of working with good-tempered colleagues are high. Befriend the difficult if you can. It makes it much harder for them to be unpleasant if you're nice to them.

It is curious to note that despite the many forms of communication available today, if you receive a personal note from someone, with a hand addressed envelope, you are almost certain to feel good about it. It is rare to get handwritten messages and what it conveys is that you — the recipient — is worthy of respect. Should the message contain an unsolicited thank you or testimonial for something you've done, you'll probably be walking on cloud nine for the rest of the day.

One of the easiest ways to disarm the opposition is to be charming and to smile. Many people have wonderful natural smiles, but due to nerves or apprehension, their faces frequently set in serious expressions. It takes far fewer face muscles to smile than it does to frown. If you are going to praise someone for something they have done, you really should be smiling when you speak to them.

A sign of a confident person is someone who while speaking in public, or giving a performance, allows themselves to project a warm encouraging expression. When a smile lights up your face, people will notice you. It is quite likely that their natural reaction will be to smile back. Imagine the powerful advantage this gives you in your potentially awkward situations. People who smile give the impression of being pleasant, attractive, sincere and confident. It relaxes those around you, with whom you are about to communicate.

Charm, good manners, politeness — these are all somewhat traditional standards. Do you notice a person's eyes when you are with them? Do you find that their attention remains focused on you throughout the exchange? Or do their eyes stray when someone walks past the door, or if a commotion takes place outside? Keeping your eyes and ears directed towards the person you're with is vital if you are trying to placate them or restore calm to a difficult situation. By creating the impression that your attention is all theirs, you will have a strong effect on the bridge building process in your area of concern or conflict. All these things build confidence. A conflict can be resolved or a relationship enhanced purely through a display of confidence. Self-belief and self-assurance are vital if you are to realise your potential and maximise your success at dealing with challenging individuals.

Top tip for property professionals

Confidence — like a muscle — needs to be exercised if it is to develop.

One of the nicest things about praise is that it stops the majority of people complaining. Whingeing is becoming increasingly common at work. If there are one or two Olympic standard moaners among your professional colleagues they need to be managed properly to move them away from their aggrieved attitude. Simply admonishing them and suggesting they 'keep a still upper lip and get on with it' won't work. Frustrated colleagues are not always wrong, or being difficult. Sometimes there are fundamental problems and these should be investigated swiftly by your project manager.

Negative perspectives can be dangerous and if there are a number of people making waves within a team, this is where the rot can set in. You cannot afford to have a vicious circle of doom-laden prophets just fuelling their own gloomy forebodings. This sort of enemy within is insidious and can, like a virus, lie in wait until another unsuspecting prey comes along ready to be infected. One of the most effective antidotes is praise and reward of even the smallest amount of success or progress. This will inevitably lead to bigger things. Why not, for example, keep a 'success diary or chart' so that project team members know when they have contributed to a mention in the book or on the wall chart.

Top tip for property professionals

Merit is often an obstacle to fortune, the reason is it produces two bad effects — envy and fear.

Proverb

Dealing with professional jealousy

There is a huge amount of professional jealousy reported in the workplace and it is an issue that can badly de-stabilise a project team and requires careful handling. You can't simply charm someone while they are suffering a particularly violent attack of the green-eyed monster. If you're doing really well someone could be experiencing envy and resentment about you. Success sometimes spawns jealousy among professional colleagues. One reason for professional jealousy is that of feeling threatened. Is someone feeling resentful towards you because of your recent promotion? Or are you popular and polished and they are not?

Some people regard work as a competition with only one winner. Should you be seen to be edging closer to the chequered flag, they may see you as the winner and them as losing. The only way they can deal with this is to push themselves forward and hold you back at the same time. Job jealousy can take many forms — from the odd 'forgotten' message, or mislaid written instruction to a formal request for information which fails to get delivered. Should you feel that your work is being undermined or criticised unfairly you need to make a note of such things. If the behaviour becomes personal, remarks about your appearance or an aspect of your private life, you should make a move to counter it.

Developing a thick skin helps, as this will afford some protection from your work-place monster. Focus on what is important and ignore what is trivial. Don't allow yourself to openly show pride in your promotion at work. If you want to bask in self-congratulation, make sure you do it out of the work place among friends. You need to maintain your own standards, rather than dropping them to the jealous colleague's level. This will do you a lot of good and you will be confident you are not contributing to making the situation worse.

You may have to be brave and tackle the person out in the open. Confront him/her and ask if they have a moment, because you have something on your mind which needs to be discussed. You could say it won't take long. Your request for a chat will probably be refused, in which case ask whether this is just a bad time, and would they suggest a better time? Don't be put off by them, persist until they agree to talk to you. You could open the conversation by saying that you feel there is a bit of tension between you and you wonder if everything is all right. If they reply along the lines of 'I've no idea what you're talking about' then you can say, 'Oh good, I must have been mistaken. I'm glad because I want us to be able to work together.' Make it clear that should they have something they wish to discuss in future, they must say so. It is important that the exchange concludes on a positive note, where both parties can exit with some dignity.

Should matters not improve, you will have to face another one-to-one at which you become more assertive. You will need to spell out the point of grievance — 'it is not acceptable when you do ... I find it difficult working with you. Please make sure that in future you do ... Are you okay with this?' Unfortunately if this doesn't work you have no choice but to make a complaint along formal lines. Disciplinary action may follow but since there is little chance of you establishing trust and rapport with this work colleague, there is nothing lost.

Whatever you do, please stay professional. The most important thing is to try to work through this difficulty without raising problems with others in the department.

Matching technology to purpose

9

If the human race wants to go to hell in a basket, technology can help it get there by jet.

Professor Charles M Allen

This chapter is out of date: even as it is being written the pace of change means that by the time it is in print and you read it, things will have moved on. This does not matter. What is important is the attitude to *information technology* and its effect on communication with other property professionals. Today there are more ways to communicate than ever before. As little as 25 years ago there were only a few different forms of communication commonly used. If the matter was urgent, you went to see someone or sent a telegram. If it was less urgent, you telephoned them. If it didn't require an instant response, you wrote them a letter and if it was not important, you probably forgot about it entirely.

There were three main methods of communication, oral direct, oral indirect and verbal. Exchanges of information by letter usually took several days, but no one seemed to mind or thought it particularly worrying. If you needed an instant response, you sent a messenger or courier, who hand delivered your communication, waited for a reply and brought it back to you. Records were kept manually and information generation and retrieval was much slower. Expectations were not so high and time-limits were far more generous than they are today.

Now there are all kinds of different and exciting ways to communicate with someone, whether they are sitting at the desk next to you or on the other side of the world. Computers have brought us email, voice mail, voice pagers, conference calls, teleconferencing, faxes, mobile phones, text messages, electronic messaging services, satellite links and so on.

Today's busy property professionals communicate far more than ever before. It is less formal and non-hierarchical, fast and furious. Formal communication still has its place, in contracts, agreements with contractors, consultants and other legally binding documents. Emails and faxed documents are now regarded as legally binding which is something most property professionals would do well to be aware of. Think of recent press reports where email paper trails have led to some unfortunate's downfall. Expectations are far higher now, people are less patient than before. Swift responses to communications are required and demanded in many situations. Computers heralded the *'paperless office'* but most people's desks seem to contain as much paper now as ever. They have also given everyone a far greater number of ways to waste time than ever before. How many times have you failed to obtain a piece of standard information because of computer failure? If you hear the message 'sorry, our computers are down and we cannot access any information at the moment', do you bless or curse technology?

What about if your company's computer system develops a virus? This could cause the whole network to grind to a half while it is de-bugged. How many millions of pounds do worms, Trojan horses and similar viruses cost industry per year? Most organisations depend on their computer technology to enable them to operate. It is normal practice today to equip every member of staff with the latest model of computer so that each person in the firm can send and receive email. If you are not harnessed to the company's computer network you are likely to be marginalised. Who do you know who doesn't have an email address?

Top tip for property professionals

Technology ... the knack of so arranging the world that we don't have to experience it.

Max Frisch (1911–1991) Swiss Playwright

Process automation

Before the emergence of computers business processes were manual ones. Accounting systems, payroll, staff records and sales would have been kept entirely by hand with the assistance of an adding machine. Work which took hours, day or weeks can now be accomplished in a fraction of the time. Some other processes, such as stock control, customer information and sales enquiries would also have been carried out with the help of manual systems and are now commonly automated. Most people now have personal organisers, calendars and planners and these are built in to their computer systems. Paper based diaries may never disappear entirely but they are nowhere near as powerful as their electronic successors. Who wants to refer to a desk diary when you can, at the tap of a screen, locate suppliers and clients' names and addresses and details, schedule meetings, track projects, analyse numbers and enable a reminder and alarm system to ensure nothing is forgotten.

What are you gaining?

The management of information predominates professional life. The sooner you obtain information, the faster you can use it. The more efficient the means of accessing the information is, the greater the edge you have over competitors. Improving the management of information is the main reason why companies invest such huge amounts of money each year in their information technology. They commit a large proportion of their annual resources to buy and upgrade their computers, install improved voicemail and email systems, training budgets are increased to help employees obtain and improve the skills they need to use these new tools of the electronic and technological age.

But has all this expenditure made companies and their employees more productive? Is it measurable in terms of profits and other gains? Is there a clear correlation between the implementation of office automation and increased efficiency? By installing computers and other information technology a company does not automatically gain in staff efficiency and improved productivity. Every responsible professional knows that it takes times to improve work processes. By investing in new technology hastily and without upgrading and improving processes the net result is things can simply go wrong faster.

You no longer need to be in the office all the time to communicate with your clients, colleagues or staff. You can be anywhere and no longer is the 9–5 office culture operational, business is becoming a 24/7/365 affair. If you know that you can be contacted at any hour of the day or night, it could be argued that technology is not always advantageous. In extreme cases it plays havoc with many professional people's work/life balance.

Top tip for property professionals

IT is the most *valuable* and *vital resource* in today's world. It effects the *management of people*, the *management of activities* and the *management of information*. Of the three the last is the most critical. Used correctly IT can make dealings between professionals more efficient. Used incorrectly, it makes communication less efficient faster — it can even accelerate bad practice.

IT: a vital resource

Organisations who invest in the latest technology can work twice as fast and you can be anywhere, anytime, you want. Whether you want to organise meetings with colleagues or clients in five different countries, confirm their attendance, enter the meetings into your electronic diary, check your notes and be reminded half an hour before each meeting, this is all possible at the touch of a key with your electronic organiser or by blue-tooth technology on your hand-held computer.

Some organisations are electronically linked to smaller ones so that they can outsource certain functions. Property professionals who harness such techniques can take advantage of new opportunities if they are highly skilled and in niche markets. Inexpensive computer aided design and software enable work to be delegated swiftly and economically.

One of the other benefits of such rapid pace of change in IT is that groups of small companies can easily use information links to form 'virtual corporations', gaining market share while enabling each to concentrate on its core skills. Mobile computing allows companies to compete around the world without setting up expensive branch offices. 'Virtual' offices can spring up anywhere.

Businesses that harness ICT (Information and Computer Technology) will optimise the return on such investment in its computer system in a number of ways, because it can be used to:

1   help generate more income
2   stop money spilling out

3 help achieve better performance from people
4 make better use of physical resources
5 improve productive processes
6 improve money management
7 help deliver strong intangibles.

There is a business premise which has been used successfully by a number of companies to increase effectiveness and productivity. This is applicable here with regard to the use of the IT. It is known as 'the 1% theory'. If you can achieve just a 1% improvement across 15 factors which contribute to business performance, the multiplier effect is considerably greater than the sum. Working on three years' business performance figures, a multinational company with sales turnover in excess of £2 billion, applying such a model could improve a £167 million profit figure by a further £83 million. Few people would argue that increasing the efficiency of any operation by 1% is impossible. This proves an excellent starting point when considering the advantages of harnessing IT for increased business efficiency.

Some companies have a tendency to become *over-absorbed in the IT processes* they are operating. This is a common failing of over-doing a task. The European Business Excellence Model (IFQM) easily illustrates this.

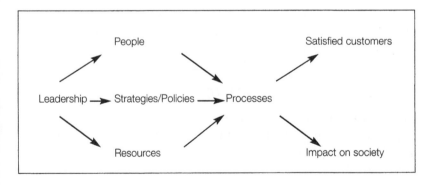

Logistically there are still a number of problems to overcome, for example in India there are 16 main languages. If your organisation deals regularly with this country because of outsourcing, it will perhaps be a familiar problem. Software is currently being developed to enable translation between those languages in order that communication can flow more easily across that country alone. The complexity is compounded when you consider the rest of the world, China, South East Asia, South America and Africa where there are hundreds of other languages.

Top tip for property professionals

Technology does not drive change ... it enables change.

Source unknown

Benefits of technology

Saving money

IT saves money by improving financial and operational reporting, managing overheads, expenses policy and administration, time costs, material costs, purchasing efficiency, asset management, budgeting, wastage of materials and resources, control systems.

Enhances performance from people

IT enhances performance from people through knowledge sharing and acquisition, communication tools and processes, standards and processes, skills and other training, presentation aids, incentive programmes, time management.

Improves use of physical resources

IT enhances the use of physical resources by improving deployment, capacity and utilisation, maintenance, replacement scheduling, purchasing, inspection, documentation, measurement, improvement programmes.

Increases productive processes

IT increases productive processes by improving scheduling, capacity planning, workflow, system design, quality, competence, measurement, materials, logistics and distribution, back-up, monitoring, change management, project management.

Assists money management

IT assists money management by increasing product/service performance, warranty, billing efficiency, investments, purchasing efficiency, work in progress management, currency.

Helps deliver strong intangibles

IT helps deliver strong intangibles such as customer relations, perception of quality and value, communication, goodwill, customer-centricity, brand recognition and equity, learning organisation, values, integrity.

Top tip for property professionals

The true value of information and communications technologies is that they enable better communication whether internal or external, and help organisations operate at a higher level of efficiency and profit.

The issue for many people, given access to so much data and technology, is how to distinguish what will really make a difference. The business benefits gained from applying appropriate investment in IT can increase standards and delivery of the foregoing processes. The benefits are all persuasive. These models can be improved and enhanced and the practical impact of correct communication harnessed to IT is measurable.

Email use, style and culture

Email and electronic communication is one of the quickest ways of communication with others on a network. There are millions of computers sending emails at all hours of the day and night and every recipient has a unique address. There are several types of organisations in this country and it is possible to work out from the email address what type it is: for example

ac	academic organisation
co or com	commercial organisation
edu	educational establishment

gov	government body or agency
net	a company running the Net
org	a not for profit organisation

In the working environment, emailing is often used as a *management substitute*. In its most exaggerated form it cuts down the need for face-to-face communication. This can be vastly *cost effective*. It is possible to conduct meetings, correspond with the whole world and use voice and visual contact without leaving your desk. Because it is *rapid* and instant, it brings pressure on the individual responsible for creating the email to *'get it right' first time*.

The attractiveness of emailing is without doubt its *speed and volume*. It is almost *instantaneous*. Mail is sent immediately you click the 'Send' button. Your message should be received very quickly after it is sent. The speed of their reply is dependent only on how quickly they check their email in-box. When used by service organisations, emails have the advantage of being a *precise method* of communicating one to many. As a process it is attractive because the expectation is *acceleration of information exchange*, service or sale of product.

Email is *less formal* than writing a letter. The main purpose is not for lengthy communications but for *short, direct information* giving or gathering. Lengthy emails are difficult to read and absorb on screen. Some people are quite relaxed about the occasional odd spelling or misuse of punctuation. But if you are able to proof read fairly quickly it helps avoid misunderstandings when being read by the recipient. You can of course use *spell and grammar check systems* before sending emails.

When *replying to an email*, you don't have to worry about finding the sender's name and address and job title. Replying involves the pressing of a button, their address appears on the top left hand box of the reply page. It is possible to keep the copy of their sent message on the page, so that you can refer to it when replying. You can communicate across the globe and its various time zones within minutes for the *cost of a local phone call*.

There are of course *hidden costs* but these are mostly connected with what goes wrong, and the *time* involved in *putting things right*. When changing systems, it is normally necessary to intervene manually and the IT department controls this in most companies. If there is no in-house IT department, this aspect of business is outsourced to expert sub-contractors.

Junk email or *spam* is just as irritating as the junk mail that arrives through the letterbox. It can be controlled by using on-line devices offered by the ISP. Otherwise the responsibility rests with the user. Caution should be exercised in opening emails from unknown recipients. *Viruses*, Trojan horses and worms can invade the computer system if care is not taken. *Emails with attachments* are the most likely ones to carry viruses, although the sophistication among hackers is growing at an alarming rate.

Presentation, style and content

Emails are more *informal* than letters but still require certain *criteria* as regards style and content. When emailing, you are competing for attention and have nothing but words to compete with. An email should consist of the following four *elements*:

- *be brief* — use plain words
- *direct* — clear presentation avoiding ambiguity
- *logical* structure
- *response* required and if so, give a reason why.

Whether emails are being sent internally or externally, as a substitute for a letter or not, it is important to ensure these rules are observed. Emails like letters are *versatile* forms of communication. These are some of the *effects* an email can have:

- Inform   meeting arrangements
  staffing issues
  new systems or processes
  commercial changes

- Persuade request for leave
  request for more money
  selling document/letter
  proposal

- Instruct   how to use
  how to claim
  how to construct
  direct

- Record minutes
  client meeting
  time sheets
  activities
  important agreements
  appraisals.

Before sending an email, consider the following:

1 What is the *objective*: purpose of email? Do you know what you are trying to achieve? Is the email a request for information? Do you wish to have a response, if so, make this clear by indicating when a reply should be received. Are you circulating standard information which can be read by other people? If the email is a quick response to a query, make sure that what you say is correct. If you are unsure, explain that this is an acknowledgement of receipt, and you will come back to them as soon as you can. If you do not know what the objective is, think carefully before sending your communication.

2 What is the *background* to the issue? Is the reason for sending the email something that is to do with a problem in a project? Is there an explanation, excuse or apology required? Is it to elicit more information or to provide detailed answers to a query? For an email to be clearly understood, there must be a reason why you are sending it. If you don't know, check with a superior before going into print.

3 Who is the intended *recipient*? Will it reach them direct, or be read by another person? Email in-boxes are not necessarily openable only by the person named in the 'Send to' box. It is possible that other colleagues could have access to this person's mail box, due to department hierarchy, or the fact that messages can be forwarded to a colleague or assistant, when someone is sick or on holiday. It is important to bear this in mind when writing a message to save embarrassment.

4 What *style* are you using? How is it being presented? Is the style really informal? Are you reply to a message which was half encrypted with lots of missing capital letters, text message style shortened words, emoticons etc? If so, that is fine. But think carefully what impression the style of the email gives to someone

who is opening a communication from you for the first time. Politeness matters and you never get a second chance to make a first impression!

5   Choice of *content*. What is the email saying and is it being clearly communicated? There is nothing worse when you are short of time than to read an email and wonder what message the sender is trying to convey. Because email is speedy and concise, try to avoid vagueness and ambiguity. If the email covers complex matters, it is better to explain that a document follows. Emails are meant to be read quickly and the content should reflect this.

6   Is there a *conclusion*/recommendation/response required? If so, is this obvious? For clarity of purpose, it is important to make any request at the end of the email. By saying, it would be helpful if you could bring this information with you when you meet at 4pm, it gives the recipient a clear message that he has until 4pm to complete the task. Finishing off an email with a direct instruction, or repeating the purpose of the message, will leave the reader in no doubt about what your intention is.

7   What, if any, *attachments* are being sent? Specify any attachments clearly. If a device is used to 'squash' information together — such as zip files, it is always helpful to explain what system you use. If the attachments require certain software to open them, explain what is needed. This is particularly important where graphics and images are being sent. Some of these attachments can take ages to download and it is helpful to say so.

Top tips for property professionals

Putting yourself across appropriately in an email is important, because it is instant and non-retrievable. In oral communication you have not only words, there is tone of voice, facial expression, posture, body language and gestures.

Attention to detail

As email is a rapid and concise form of communication, there are some good tips for *email etiquette*. These are some of the most important points to remember

- Correct format or *house style* — this is often available as a template. Make sure it matches the style used in the company's letters and faxes and check what other aspects of layout are expected to conform.

- Typography/*font* — most companies have a prescribed font and style but the font type can be chosen from the drop down list box. The screen shot shows the font and size selected. You can also select the option for formatting the text in your email, such as bold, underline and italic.

- *Subject* — writer reference, case number or project name. This is just a polite way of ensuring that the recipient can save time by reading what the email refers to. If you are sending an email to someone about a particular matter, it is helpful if they understand immediately what the message is about.

- *Salutation* — are you on first name terms? Do you need to write in more formal style because you have not exchanged correspondence before? Do you know the name of the person to whom you are writing or would it be an impersonal salutation?

- *Punctuation* — beware ambiguity. A missing comma or no full stop can often cause confusion. It may be 'cool' to lose capitals and miss out dots and dashes but if the reader is left puzzled by the meaning, you are less likely to get a useful exchange of information.

- *Line length* — short sentences and line length make for easier reading. This is explained in more detail further on. Do not use complex sentences or syntax. Short and sweet is best.

- *Paragraphing* — options are available from the drop down list, including headings, bulleted lists, numbered lists. Paragraphing should be used where there is a change of topic or subject, so that the reader is aware that a new point is being introduced.

- *Consistency* — if the email contains numbering take care. It is extremely irritating if the numbering changes in style or is inconsistent. If you are making a number of points, stick to (a), (b), (c) or 1.i, 1.ii, 1.iii or whatever style or format you prefer.

- *Valediction* — unlike a formal letter you don't have to sign off 'yours faithfully' or 'yours sincerely', however in some cases it may be appropriate to end with an informal send off. Many people use 'kind regards', 'many thanks and best wishes' or more impersonally, 'yours'.

- *Writer details* — title, company. With emails it is possible to set up as a 'default' a signature which appears at the foot of the message. This includes your name and title as well as details of the company you represent.

- *Contact details* — continues on from the preceding point. This should include for ease of reply the best contact number for reaching originator, should a telephone call response be appropriate. In some cases it is sensible to include the full postal address of the company, with any strap line, if that appears on the company letterhead.

- *Enclosures or attachments* — as mentioned before, these should be clearly described and mentioned in the text. If they are in different format, such as PDF files, it is a good idea to ensure beforehand that the recipient's computer is able to receive these files in readable form.

Blocking unwanted messages

Anti-spam filters are recommended and are incorporated in corporate IT systems as standard. They do not catch everything, but they certainly reduce the volume of spam reaching email in boxes.

There may be a number of reasons why emails need to be blocked from particular senders. *Junk emails* are just one of the main reasons, others include people from whom you no longer wish to correspond. By setting up barriers supplied with your email package, specific email addresses will be blocked. You can easily remove a sender from the *Blocked Senders List* by selecting the address and clicking the Remove button. When a sender is blocked, their message will be diverted straight into the Deleted Items folder. Don't forget to empty this folder regularly, otherwise it can become clogged with unwanted messages.

Beware opening messages from *unidentifiable sources*, particularly *with attachments*. These can contain viruses or micro-programs that can access and send to a third party information from your computer.

Digital Signatures and other security devices

Electronic signatures are being used widely as more people send information by email. In addition, it is more important than ever that emails can't be read by anyone other than your recipient. By using *digital IDs* or signatures you can ensure that no one is pretending to be you and sending false or misleading information under your name.

Digital IDs in Outlook Express can *prove your identity* in electronic transactions. This is rather like producing your driving licence when you need to prove your identity. Digital IDs can be used to *encrypt* (code) emails to keep them from other people reading them.

Digital IDs are obtained from independent certification authorities whose websites contain a form which when completed contains your personal details, and instructions on installing the digital ID. This is used to identify emails and ensure security of your messages.

Encryption is a better way to send sensitive information by email. It is a form of *electronic code*. One code is used to encrypt the message and another code is used to decrypt it. One key is private, the other is public. The public key is passed to whoever needs to use it, whether they are sending the message (in which case they would use it for encryption) or if they are receiving the email (they would use it to decrypt the message).

Some email systems allow a note to be shown when an email has been sent, received, opened and read by the recipient. This can be important in some time-critical instances, such as in finance, banking, law and property.

Viruses

It is advisable not to open emails that may contain *viruses*. A virus is a small piece of code which is deliberately buried inside a program. Once the program is run the virus spreads and can damage the data in your files or erase information on the hard drive. Thousands of viruses exist and new ones are being invented every day. *Anti virus software* which can search out and destroy a virus is essential. These should be updated on a regular basis.

Virus checkers should be run before opening any emails with attachments. Some ISPs are now scanning emails on their way through and will inform the email account holder if any possible problem emails have been detected. You can then make an informed decision as to whether you are going to risk opening up the email or not.

Viruses are distributed over the Internet in a number of different ways. Software downloaded from the Internet itself may contain a virus, they can be transmitted by an email attachment. It pays to be suspicious of an email attachment from an unknown recipient. A macro virus is hidden inside a macro in a document, template or add-in. A document with a macro virus not only infects your computer but also other computers if you pass the document on.

Most organisations have a system which automatically backs up files and company data to safeguard against virus infection and loss such as fire. If passwords are used in your computer system, these should be changed on a regular basis to stop hackers and avoid misappropriation.

Create a filing system for emails received from regular correspondents. This is useful, tidy and saves a lot of time. Depending on your email system, this can either be done within the email program itself or by creating a new file or folder in Windows.

Spell check and other helpful tools

Spell check

Spell check is a great tool for those whose spelling is less than perfect. It not only saves time hunting for the dictionary, it also gives the option of 'learning' the word if it is one that is used often. Watch out for the wiggly red worms that underline difficult words. If you see these appearing frequently in your documents, it means there is an unusual name or spelling that needs checking. Good systems which incorporate the 'learning' facility allow you to select extraordinary spellings, such a surnames, place names and company names, which saves a great deal of time.

Be aware of easily confused words and use 'spell check' with caution. For example, how easily a sentence meaning is changed by the wrongful substitution of the word 'now' with 'not'.

Extract from a letter ... 'After further consideration I have decided that your request for a salary increase of £10,000 per annum will now be effective'. Try that sentence again substituting the word 'not'.

Grammar and language checks and the thesaurus

Grammar check/language/thesaurus are other excellent features. They show up on text with a wiggly red or green line where there is doubt as to syntax. It is possible to select alternative words or phrases to avoid confusion when using the grammar check tool. As a matter of best practice, when emailing try to use 'lean English' wherever possible. The following words and phrases are examples of exactly the same in meaning, but in its simplest form:

In the event that	if
Subsequent to	after
The possibility exists for	might
Prior to	before
In order to	to
In the region of	around/about
From time to time	occasionally
In reference to	about
It is necessary that	must
Due to the fact that	because
For the sum of	for £

Tautology

Tautologies are a definite No when emailing. Avoid them at all costs.

- contained herein
- close proximity
- comprised of
- completely full
- end result/final outcome
- future prospects
- other alternative
- new innovation
- past experience
- postpone until later

- mutually agree
- recurring habit
- initial preparation
- various different
- future plans
- free gift

Less is more

When emailing, a good idea is to get into the habit of using smaller words for bigger impact such as:

administer	give
aggregate	total
alternative	choice
anticipate	expect
articulate	explain
culmination	end
fundamental	basic
indication	sign
ineffective	weak
majority	most
necessity	need
opportunity	chance
possibility	chance
reiterate	repeat
requirement	need
subsequent	later
utilise	use

Sentence lengths and their effectiveness

The average length of your sentences affects the ease with which they will be understood. The reader's process is both to decode the sentence and understand its meaning. If the sentence is long and the meaning complex, you are asking the reader to expend too much brain energy. If you are trying to get your message across and elicit a favourable and quick response, this is counter-productive.

The effectiveness of sentence lengths has been researched as follows:

Average length	Readability	Impact
Up to 8 words	very easy	90%
11 words	fairly easy	85%
17 words	standard	75%
21 words	fairly difficult	40%
25 words	difficult	25%
29 words or more	very difficult	5%

Jargon and acronyms

Emails containing jargon, text language and acronyms (where initial letters are used to make up another word) are not the best style for emails. However because of its overriding informality it is a good idea to be familiar with those that are universally used. There are many around and new ones are springing up daily due to the popularity of text messaging. Here is a selection of some of the more common acronyms that you are likely to see in emails are:

AFAIK	as far as I know
BCNU	be seeing you
BTW	by the way
CUL8R	see you later
FYI	for your information
FAQ	frequently asked questions
TNX	thanks

It would be a good idea to learn these and any other additional ones that are commonly used within your company or profession or industry sector. Beware using them in emails which are being sent outside the firm where the recipient does not understand them. When using acronyms to external people, be courteous enough to use full terminology in parenthesis afterwards.

Attachments

Documents and files can be *attached to emails* which are useful when *speed of dispatch* is essential. Attachments can include word-processed documents, images, sound or video files. It is possible to email computer programs.

When an attachment is sent the email program copies the file from where it is located and attaches it to the message. *Image files* can take some time to upload and download, so it is advisable to keep these to a minimum if speed is of the essence.

Compress files that are being sent as email attachments. This will reduce the upload time while transmitting the information. It also speeds up the download time for you if someone sends you a large file that has already been zipped. The advantage of sending documents and files as attachments is the speed and efficiency of communications. The recipient of the documents will be able to keep these on file and can edit, return or forward them on as necessary.

If security is an issue, an attachment should be sent in *PDF format (Portable Document File)*. This format prevents the document being edited by the recipient. This is a security device and the document can be printed off but not amended. This is very safe and secure for sensitive material.

Hyperlinks

Inserting hyperlinks into email messages are particularly useful when sending information to people. Something that is available on a web page that needs to be communicated to your recipient (of which you do not have a copy), simply insert the *hyperlink* into the email message. The recipient simply clicks on the link and opens the web page.

Need to act

Emails, because of their popularity and versatility, currently threaten to obliterate all other forms of communication. It is vital to stay on top of them. Here are some good practice tips.

1   Clear your inbox every day.

2   Allocate items that are time critical and items that will require work later.

3   Delete any emails that are irrelevant or unimportant.

4   Unsubscribe from email lists assists clearing the clutter in your in box.

5   Copy yourself — (cc) or blind copy (bcc) yourself a message when responding to emails where arranging a meeting or promising a response or sending information.

6   Ration the reading of emails to three times a day, early morning, midday and end of day. Reading messages as soon as they flash on the screen causes severe interruptions.

7   Delete messages that you've dealt with and empty the deleted folder regularly.

8   When replying to messages where the original text remains on the screen, make sure that the responses you insert are in a different colour to draw attention to revisions and insertions.

Email etiquette

Suggested tips for good email etiquette include:

Don't –   send emails just because they are easy
       –   enter text IN CAPITAL LETTERS. It is taken as shouting.
       –   use them as a substitute for properly delegating a task to another
       –   send them to discharge yourself of responsibility
       –   put something in an email that is confidential, it can be abused
       –   forward someone's email without their permission
       –   assume your recipient wanted it and is desperate to receive it.

Do   –   think and use the 'send later' button if in doubt about sending an email
       –   be precise ... to eliminate follow up phone calls
       –   reply promptly — because email is quick, a reply is generally expected
       –   be polite and friendly but never assume familiarity with jargon
       –   keep attachments to a minimum
       –   avoid gobbledygook.

The following extracts were from a car hire company's email to a client

Ensure environmental and comfort systems are confirmed as being operational prior to the commencement of a journey.
Translation: Make sure the heaters and lights are working before you set off.

Ensure environmental and comfort systems are adjusted to meet changing conditions.
Translation: Turn the heaters on when it is cold and the lights when it is dark.

Ensure location factors and conditions in which manoeuvres are to occur are considered with regard to safety, minimal disruption to other road users, residents, legal constraints and regulatory requirements.
Translation: Look where you are going, check mirrors etc.

Ensure the vehicle is effectively manoeuvred to change direction.
Translation: Turn the steering wheel when you reach a bend.

Ensure awareness and anticipation of other road users in the vicinity of the manoeuvre is maintained.
Translation: Look where you're going.

What does your email technique say about you?

Actions speak louder than words, as the saying goes. But you can tell a lot about a person from their email too. An email can provide a window to your status in the workplace, work habits, stress levels and even your personality.

You may receive emails from someone who uses 'higher status' techniques. These contain a greater level of formality and tone, and lack the detail of a lower level member of staff. No-where will you see 'cheesy quotes, smiley faces (emoticons) or joke mails.

Emails are such a valuable communication tool for everyone, but if abused or used carelessly, can cause trouble. In summary, here are ten basic tips for better email technique.

1   *Use email as one channel of communication, but not your only one*
    This is important, don't be lazy just because it is fast and easy.
    Emails can document discussions and send high impact messages around the world at the click of a mouse. But it can also mislead managers into thinking they can communicate with large groups

of people through regular group emails. Use email widely but not as a management tool. It is not possible to reach everyone, and the 'impersonal' non-direct contact means that people feel can feel slighted by the loss of the personal touch.

2   *It pays to keep it short and sweet*
Emails that are longer than a full screen tend not to be read straight away. They get left till later and often not until the end of the day or the next morning. It is important to judge when it is right to put down the mouse and seek the person out for a face-to-face, or pick up the phone and speak to them. Constant email actually erodes time unnecessarily.

3   *Message clearly — cut out the codes*
Email requires clarity of purpose. Be sure your message comes across without any doubt or misunderstanding. Also it is important to be sure to whom your message needs to be addressed, and who needs a copy for information. In terms of actions and priorities, use lists or bullet points for clarity. Response buttons (or similar) should be used if you need to see who has received and read your message.

4   *To encourage open communication*
When using email, request people to respond with questions or queries if they wish. It shows that you are concerned and available to help.

5   *Don't use emails to get mad with people*
Far better to save anger for face to face encounters (where facial expression and body language can be used to great effect) or over the phone where tone of voice can speak volumes. Sarcasm, irony, criticism or venom are not appropriate when sending emails. They often come over far more harshly than intended.

6   *Humour should be used with caution*
By all means use wit and humour to lighten a heavy atmosphere but emoticons, smiley faces and joke mails are not appropriate in the work environment. If being facetious is usual for you, it may make it more difficult to strike a serious note when you need to. Joke emails are banned by some companies, as they are too risky. Too many joke emails erode your attempts to send serious ones.

7   *Suspend reaction — use the five minute rule*
It is often wise to delay sending a hastily written email for five minutes before pressing the 'send' button. If you are angry when you write something, it is a good idea either to take a break or go for a walk or do something else, before writing. Otherwise if you write the message, once you have cooled off take a moment to review it before sending it out.

8   *Set aside time to deal with emails*
Because of the growing importance of emails in terms of mode of communication, make time to deal with them. If you can't find time during working hours to read and answer your emails, adjust your tasks. You should not have to stay late at work to deal with them. Maybe you need to delegate some of your other work or pay attention to some time management techniques to improve output.

9   *If you are doubtful about writing emails effectively, get some training*
Help is available within most companies in-house training department, or consult administration staff. Do not avoid sending emails because you are unsure of how to do it correctly. Ask for help. Anyone with whom you communicate may judge you on your emails and you will want to appear efficient.

10   *Take advantage of tools such as spell check and thesaurus*
To avoid errors and complicated sentences, use the tools provided to ensure clarity of communication.

Fit for purpose

Top tip for property professionals

To become an experienced and influential networker requires relationship building skills. To develop these, you need a strategy as well as persistence.

You may have thought that networking was a rather peripheral skill for property professionals but it is a valuable asset. If you are going to scale the ladders on your career path, you will do so far more effectively if you add a bit of structure and purpose to your relationship building skills. In addition to your professional qualifications, the ability to relate to other people is important. A previous chapter has illustrated how you could make a start on the networking process if you are coming to it for the first time, and what types of networking events you might find. Creating the right impression and communication styles and techniques has also been covered.

Networking strategy

This chapter will show how to add structure and substance to what will become your personal networking strategy. Having a strategy — whatever the objective — makes progress easier. There's no point in doing something if you don't know why you're there. It helps to have a purpose or a plan. Starting with the end in mind is always a good idea and that is something you'll need to consider. So before going any further give some thought to what you as a property professional want your networking to achieve. Have I mentioned that building successful personal or professional networks is a lengthy process? Networking does bring rewards — but not quickly. So try not to be impatient. Rushing into things is counter-productive. You may have seriously underestimated the length of time your network will take to mature. Good networking relationships develop best when the people involved think carefully about what they are getting into and come to a considered decision.

What about your oldest friends? Are they also your closest confidantes? If the answer is yes, perhaps it's because they have known you for such a long time. Did you grow up together? If so, you'll have lots of things in common. Maybe your parents and families were friends before you were born. Were you in the same class at school — spending time together every day for a period of several years? Or perhaps you went through university or college at the same time? In that case you probably have many colourful memories of your student days.

Top tip for property professionals

It's time to set out your networking objectives, understand and document your goals.

Building on your existing network

Have a look at who is already in your data base, and who you've met recently on a professional basis. Do any of these people seem possible candidates for building connections with? Think about it and try to make a list. If you are wondering what sort of list, consider the following.

One way to begin is by identifying common links and themes to each and everyone in your data base. If you adopt an analytical approach it might help. Have you done any mind-mapping? On a sheet of paper start with you, right there, in the middle. Branching out from you, in a number of different directions, are the routes that take you to different groups of people in your life. These could include:

- work internal (the people who work with you, colleagues, boss and staff)
- work external (the people you meet through work but outside the office clients, suppliers and others)
- leisure (those you know from your other work-related activities—people you meet at quasi-social functions).

Now, like satellites spreading out from a central point, write the names of no more than ten people under each group. When you've done that, have a look at the paper and see whether any of these people could be linked together, other than through the one main connection point — which of course is that each of them knows you.

This requires a bit of lateral thinking but here goes.

One of your members of staff has a daughter who is looking for work experience in her gap year. She is interested in building design and you work with a number of different companies who take on students for short work placements. If you mention that there is a young person available who is reliable and seeking work during the summer months you could put your work colleague in touch with several people who could possibly offer his daughter some temporary work.

You can, from your knowledge about your contacts draw links and connections between them. There is not always a 'quick fix' result to connecting, but by keeping an open mind — ie a friend is looking for interim accommodation because he's just sold his flat and doesn't want to buy something else immediately — you may hear something helpful from a completely different direction in the next few weeks that could make a difference to him.

Top tip for property professionals

Never refuse any advance of friendship, for if nine out of ten bring you nothing, one alone may repay you.

Madame de Tencin

Building a database

In case you haven't already got a data base — you should think about making one, or updating the existing one if it is out of date. The sort of information you need to record when setting up your strategic network of contacts includes

- name, address, telephone numbers, fax, mobile, email, website
- employer's name and job title (if applicable)
- classification of contact (colleague, client, service provider)
- type and frequency of contact — ie weekly/daily — work/ monthly — business meeting or social such as a dining club
- personal details — birthday/family/hobbies — where known
- background information — (just moved back from UK/at Uni together)
- any existing links and mutual acquaintances.

Who really is who?

While you're doing this, have a look at the structure of your network of contacts. Are they categorised correctly so that it is possible to access people quickly? Do you have enough groups, categories and sub-sections?

Friend, Family, Other is okay but it won't really be sufficiently detailed. You should include whatever groups make it easy for you to find them when you need. Say you're attending the wedding of a cousin in New Zealand next Spring and you want to know if you've any contacts in the area who you could visit while you are out there, could you find out, at the press of a button, if you know anyone in that geographical region? If you have set up your data base so you can do that, spend time organising it properly. It will pay dividends in the long term. Create whatever fields necessary so that you have the information you might want to access at the press of a button. Do you keep the notes section updated regularly?

Top tip for property professionals

The more systematically you keep your contacts and if you can keep it neat, clean and tidy, the more user friendly it becomes. This will make it a valuable accessory.

Set yourself a monthly reminder note to spend time updating your contacts data base. If you think it would be helpful, start lists of people you'd like to contact if you haven't seen or spoken to them in a while — say the last three months. They will appreciate you keeping in touch. They won't appreciate hearing from you if all you ever ask for is a donation for your pet charity, or inviting them to buy prize draw tickets for your old school fundraiser. People are often quite amazed if you get in touch without any particular reason or ulterior motive attached. It's one of the sure ways of eliciting 'free' information or help. Don't ask for anything — just listen and see what you're given.

The purpose about becoming known as a 'giver' not a 'taker' is that you will build long term, trusting and respectful relationships. This does require time — rush it and you will be disappointed. Developing connections with likeminded people is the aim. Why is this? Because if you have an easy and open relationship with someone, they are most likely to be able to offer the right advice or support should you need it and vice versa. One of the best ways to start off is by helping them as and when you can, or by offering your time, skills or talents should the need arise.

Top tip for property professionals

The first step to getting the things you want out of life is this: Decide what you want.

Ben Stein

Attitudes and approaches

When you start thinking about how to implement your networking strategy, you will have to consider what types of people you're adding to your contacts, how you wish to deal with them and whether there is any particular purpose. Some people, as you know very well, resist overtures of friendship and are harder to build relationships with. If you are new to the process, to begin with have a look for some friendly open characters who are fairly relaxed about meeting people and are not likely to freeze you out if you come up with some ideas for making connections with other people.

Assertive — warm characters

Expect a friendly approach from these open minded individuals. They are quite secure and confident and will be flexible if you make suggestions. They will probably quite quickly see the reason behind your initiative and support you in your endeavours. As long as they understand the purpose, that they can help someone while potentially gaining something themselves, they will be pleased. They take a personal interest in everyone they know and if you are in their network, they will have up to date knowledge about you and your work.

Accommodating — warm

You can expect a warm welcome, but so can everyone else. Their warmth does not indicate that you are particularly special. They are a step removed from the enthusiastic assertive/warm people. Allow them to express their feelings but do not let them deflect you from your purpose. These people will require more convincing than the others but tell them how and why you would like to progress the relationship, in a professional way. Make your purpose clear, including your role and theirs. Keep the tone conversational and flexible. Position yourself as a friendly contact with a willingness to help them. Then sit back and wait. They'll probably be a useful addition to your network but unlike the first category, they may send you an email, or get someone else to deal with the enquiry rather than dealing with it personally. But their response will most likely be a positive one.

Accommodating — cold

These people are a degree or two colder than the others. The best way to engage them is by letting them take the lead. Demonstrate that you know what you are doing when you initiate the contact. As it is initiated in a professional context, don't be alarmed if they sit quietly, listening, note taking and asking concise, factual and open questions. These enquiries will help them to decide whether they are prepared to get involved in the process. Be positive and polite but persistent. They are unlikely to give you an answer straight away. If for instance you were seeking sponsorship from them individually or their company for a charity run, they may say they are prepared to consider it but not for this year, because their sponsorship efforts are already taken. If they suggest you write to them in six months time, do so. They will remember that they invited you to do so and will expect you to take them at their word.

Assertive — cold

These people trust very few people, and don't like someone approaching them out of the blue with a 'bright idea'. They are introverted and do not have an open mind where relationships are concerned. They will probably be quite alarmed that they are in anyone's network anyway. They prefer to remain aloof and it will not be easy to penetrate their reserve. Say you were attempting to fund raise for a really worthy cause, do not expect a warm welcome when you meet them. Their negative attitude is not personal, they use it as a shield. Small talk should be kept to an absolute minimum. Emphasise that you are talking to them for a good reason, make your remarks short and to the point.

It could take a while to get any positive signs from them — but this is where the networking strategy helps. In time you may find yourself introduced to someone who could be an influential persuader and help you make progress with them.

Top tip for property professionals

Once you start using your data base on a regular basis to assist your networking activity, make sure you keep up with any changes of details.

Updating your network

Networks are organic and changes occur frequently. People move to new places, get new jobs, marry, divorce, produce children, go travelling. Updating changes should be done little and often, so that it won't become a mammoth task. There's nothing worse than knowing you have several months of accumulated business cards sitting in a pile in your drawer and very little recollection of where you met these people, let alone whether you got on with them or not. Some people use

Christmas as a time to update their networks but once a year isn't really often enough. How embarrassing if you send a card to someone who six months earlier got married and you hadn't heard nor have any idea who their partner is. Ideally every month is the best plan. This will lessen the risk of you shelving the task because it has become too time consuming. You will probably have sufficient recall that you will remember the meetings or events you attended, have kept the business cards or emails received if someone wrote to you as a result of making contact.

Spending time on a data base cleanse and update not only refreshes your memory as to who and what's in it, but it helps you work out which contacts you already have a connection with (even if you've done nothing to promote it). From here you can begin to create a relationship building strategy.

If you put into practice the foregoing suggestions, you should be building and strengthening an effective and valuable collection of contacts which is regularly refreshed and added to. Keep lines of communication clean and clear and use your network to develop new and exciting lines of approach. It is better to have a smaller and more manageable collection of contacts than something large, unwieldy and inaccurate.

Link minded

What are you looking to achieve from your professional network and how can you put your contacts to work? Are you planning to network for professional reasons — to promote your company's services or products, or for possible career progression? Who do you know who would be in a position to help you? Who likes doing things and making a difference? Think laterally as well as upwards and downwards. Anyone serious about networking has to be focused on what they want to achieve.

Apart from your work colleagues, clients, suppliers etc, your personal network can also include those with whom you used to work, former colleagues, past employees, ex-employers. Beyond that it reaches members of your clubs, professional associations, associated businesses and other networks.

How you make use of these connections depends on what you want to achieve and what activities you pursue.

> **Top tip for property professionals**
>
> **Be selective. It's impossible to keep in touch with everyone you meet, and it's not necessary to do so.**

There comes a time when people move on and you lose touch. But with good organisation and an effective network you can often maintain a link with these contacts through mutual third parties. There are numerous websites now that offer networked links from one group of professionals to others. This is a fairly quick, effective and safe way of increasing your contacts. It's proof positive of the rule of 'Six Degrees of Separation'.

If you don't know what this is, the idea of Six Degrees of Separation was devised by the American social psychologist, Stanley Milgram in the 1960s. His original experiment, using posted letters forwarded from person to person, suggests most people are connected to each other through a chain of about six acquaintances. (The 1993 film *Six Degrees of Separation*, starring Donald Sutherland and Will Smith, was inspired by this research.)

The small world phenomenon is real — scientists are researching to provide answers to such questions as how ideas spread, why fashions come and go, how a small failure can cause catastrophic global consequences (either scientifically or financially). The fact that the 'Six Degrees' rule works helps explain why among many things, complete strangers often turn out to have mutual friends and why gossip can spread so quickly. Think how often you read about an email (which was not intended to be circulated globally) has been sent across the world within hours, causing acute embarrassment to the unfortunate sender.

> **Top tip for property professionals**
>
> **Keep motivated. If you lose interest in your networking activities, everything falls apart. It has not been going long enough to survive being neglected.**

Apathy is the biggest threat to your networking activity. If indifference prevails, nothing will happen. There has to be an incentive to continue to build your network, increase your contacts base and develop your personal relationships. This comes down to your own personal attitude. Be active and make sure you attend regular events, receptions, parties, professional interest group workshops, private social functions. Keep up your sport and leisure activities as they can be very useful. Of course no-one will mind if you miss occasional meeting or outing. Everyone accepts that there are conflicts of priority, illness, holidays, work commitments etc. But if you join a new club or group unless you put in some initial effort and turn up to events so that your face becomes known and familiar, you will have an uphill struggle to build trust and be taken seriously. There is limitless opportunity in the property profession to meet people and make good connections. You know technology enables the circulation of information to lots of people instantaneously. But you need persistence if the relationship is to flourish. Whatever you do be alert and open to making connections and be patient and persistent in your actions.

Top tip for property professionals

The difference between 'networking' and 'connecting' is that there is more than one common thread running throughout.

Coincidences do occur and it is encouraging when they happen. Recently a colleague of mine invited me to attend a reception. Unfortunately on that particular evening, illness prevented her from attending, but she suggested I went along anyway. I did what a lot of people dread, and walked into a room full of strangers. The first person who spoke to me when I arrived was (like everyone else in the room) totally unknown to me. To break the ice, I asked him what he did. He said he'd retired last year from a career in teaching. I enquired where he'd taught and he named a few schools, one of which I recognised. I told him that a good friend of mine was a teacher there. He was amazed when I told him my friend's name. These two men had been colleagues for over 25 years and each had been 'best men' at the other's wedding. It *is* a small world.

> **Top tip for property professionals**
>
> **Do something. If it doesn't work, do something else. No idea is too crazy.**
>
> *Jim Hightower (The New York Times March 9 1986)*

Reviewing the network process

Let's review and reinforce some of the benefits of the networking process.

* *Networking* is the essential *first* stage. It is the framework or skeleton. It gets you out and enables you to meet people who can be added to your data base of contacts.

* Making *connections* is the *second* step, putting flesh on to the bone and developing the framework. Identifying the links and patterns that emerge — mutual friends, connection points such as similar professions, past links — attending same school or university.

* The *third* and final stage, *building relationships*, for a particular purpose — ie you are new to your job and the neighbourhood. You will want to develop friendships so that when you are not at work you have some social life outside the office.

Building relationships

To enable personal and professional relationships to develop from your network, you will need to show and earn respect from others. It takes time to earn respect from others, be sure to give it in return. How do you define respect? In a business context you might, for example when dealing with a new boss, bear in mind the following points.

* he is important and should be treated as such
* his opinions and position are already respected by others
* he should be dealt with as a person as well as your boss
* the benefits that are available to him by developing a good relationship with you

- what you know about him/what he is likely to know about you
- any snags that may need looking at (and there usually are some)
- what compromises will be required on your part, if not his
- how the relationship can work.

All of the above applies equally in your non-working life but it perhaps is worth concentrating extra hard where your career is concerned. Respect for other people makes relationships possible. You need it to be reciprocal if it is going to be effective and have a positive benefit.

In order to build respect between two parties, you may have to cover issues which are not always readily acceptable to the other side. It helps considerably if you can develop 'persuasive' relationship building skills.

As an effective persuader your approach needs to be seen by the people in your network as

- understandable
- attractive
- convincing.

None of these on their own is enough to secure a brilliant personal relationship. They need to be strong enough jointly to command respect from the other party so that you can be assured of a positive reaction whenever you meet. To be convincing and effective, your approach to strategic networking must be individually tailored to the other party involved in the process.

Each personal contact will need to be approached in a way that shows them you respect their point of view, aligns with their personality and interests. This will generate an immediate and positive response and interest. Bearing this in mind will get you off along the right track. It should quickly show you the reason why the identification and preparation stages are so important.

If handled smoothly, your networking strategy will appear well thought out and relevant — because that's exactly what it is. Your contact will know that he is dealing with someone who takes time and care over each contact he makes and treats each one with respect.

- *Make what you say attractive*
  Communicate clearly, make your words attractive so that the people in your personal contacts data base want to listen to you when you get in touch with them. They may be as keen to develop the relationship as you are. But how are you to know?

- *Talk about the benefits of your friendship*
  People do not encourage relationships unless they trust or respect the other party. Whatever the purpose behind making the contact — such as in your career building — it is essential that you are clear as to your intentions. Whether or not it is appropriate to share this with them at the first opportunity is difficult to say — each case will be unique. Be sensitive to their feelings and take things forward steadily. It is important to get this right. Each contact is different and in some cases you may be trying to persuade more than one person of the advantages of building an alliance. For example, you could be required to influence a group of directors, or partners regarding a possible future working relationship with another professional service firm.

- *Communicate with everyone, respect their opinion, value their contribution, and gain from the experience*
  In defining persuasion, it is important to make what you say to your personal contact credible. Most people who are experienced in professional dealings have a healthy degree of scepticism. They can be forgiven for thinking that you have a vested interest and will be looking for an element of 'proof'. The main form of evidence has to be the persuasiveness of the case put forward, harnessed to tangible personal benefits, followed by proof positive that it can be done.

- *Adding value*
  You may be able to strengthen your power to persuade people in your personal network if you have elements of added value that you can bring to the relationship building process. This will depend very much on your individual experience or position. If for example you are looking for support of a charitable venture, if you can offer:

  - good profiling of them or their company
  - excellent personal or professional publicity
  - unexpected benefits such as meeting influential new people.

  This could have irresistible appeal to your personal contact. You could, for instance, offer some other 'pro bono' service — such as helping them with one of their events.

- *Positive effects*
  Any such device as this can act in a number of different ways to

  - help you get a better hearing
  - help improve the weight of the case you can present
  - can persuade people to act now rather than later.

  However you choose to persuade people of the benefits of your mutual personal relationship, this can only be part of your organised strategic networking process. Some suggestions will work better than others. You may be able to control parts, but you will not be able to control all of it.

- *Use humour to improve creativity and lower stress*
  In your dealings with other people, the ability to use humour can work wonders. It aids communication, establishes empathy, and diffuses awkward situations.

Top tip for property professionals

A sense of humour is part of the art of leadership, of getting along with people, of getting things done.
Dwight D Eisenhower (1890–1969)

Studies show that humour has a beneficial effect on people because it raises the immune system's activity, decreases stress hormones which constrict blood vessels and increases the antibody immunoglobulin A. If you use or experience positive humour the whole brain is involved, not just one side. The result is better coordination between both sides. This means you are more relaxed, your blood pressure and heart rate are lowered and you are able to think more clearly. If that is what happens to you, imagine what a positive effect it will have on those personal contacts.

It is important to stress here that it is positive, not negative humour that works wonders. Sarcasm, irony, insults and black humour are not helpful. Avoid all forms of sexist, racist, crude and mean-spirited remarks. They work in reverse. When you are trying to woo your personal contacts in an effort to persuade them to do something you want, there are plenty of opportunities to show an appropriate sense of humour during networking events or meetings. You can often get an impression of whether humour is appreciated by looking around you. If you are in someone's office, for instance, are there any amusing signs, cartoons, slogans or pictures? Do other people seem relaxed and able to joke with each other? If you are with a group of people in a social setting, is the atmosphere jovial and light hearted?

When people have to decide whether to work or collaborate with you or not, they will be influenced by how they feel about you. By including humour in your dealings with other people, you are encouraging them to like you. Laughter reduces stress because it is relaxing and calming. It has been shown in hospitals that patients who have had 'humour therapy' recover quicker from illnesses or surgery than those who do not laugh.

If you are trying to build rapport in your organisation, or with new people you've just met outside your company but related to work — if you can genuinely make people laugh you know you have the ability to forge strong relationships. How many times have you noticed that when you are stressed you fumble, drop things or make mistakes. It is a myth that laughter is trivial. On the contrary — it is very powerful. Even just smiling can be healing and reassuring.

On one occasion I was attending an important job interview. I was desperate to be successful because it was a position I really wanted. The interviewer was a rather serious man who remarked, when looking at my CV, that my first job had been in The House of Lords. 'Yes, that's right,' I responded, 'and I've been working my way steadily downwards ever since.' 'In that case you won't fit in here,' was his swift reply. He hadn't appreciated the joke and needless to say, I didn't get the job. I am quite happy to laugh at myself — and am never short of opportunities to do so — but I learned from experience that there was a situation where I had misjudged the effect of using self-deprecating remarks.

Part 3

Keeping Other Property Professionals Happy

Getting on in the world of work

Meetings are indispensable when you don't want to do anything.
John Kenneth Galbraith (1908–) Economist

Networking and meeting other people inside and outside the work environment are important to your success as a property professional. Meetings are the most common form of professional communication and should they be successful there is nothing more effective. The problem is, despite the frequency of meetings you attend as part of your professional career, they are not everyone's favourite activity.

Most property professionals complain about the number of times they are expected to attend meetings — but should someone learn that a meeting has been called and taken place and they are not on the circulation list, they suffer pangs of mortification and anxiety. Why? Why are they not invited to attend? What exchanges of information have they missed? Was it because of something they had done — or even worse — not done? Meetings themselves can be the cause of ambivalent feelings. People say they hate them in one breath, yet the moment they find they've been left out of a meeting group they immediately wish they'd been present.

Meetings can be extremely hard work, often boring and can end up serving no useful purpose. Whether informal or formal, meetings should always serve a purpose — to advance a topic, disseminate information or prompt a decision. So much of your time as a property professional is spent in meetings — this means there is a high cost element. No one will argue that they are an unavoidable part of organisational communication, consultation, debate and decision making. They are necessary — but not all of them surely?

Top tip for property professionals

There is a saying 'the ideal meeting is between two people — one of whom is absent'.

Meeting fatigue

Are you spending so much time in meetings that you have no time to actually do any work? Do you measure your day by two-hour appointments that invariably run into each other? Is your briefcase packed full of minutes to be read and action lists to distribute? If the answer to these questions is 'yes', then you are suffering from meeting fatigue. This is a widespread phenomenon in the property and construction industry and can lead to all sorts of frustrations.

- there's 'voicemail rage', where the caller is so incensed at not being able to speak to a real person that the machine takes a battering of abusive language

- 'minute-takers syndrome' — the tell-tale laptop bag slouch and ability to type and talk at the same time

- 'speed-read disease', where minutes are read so swiftly that key actions are missed with occasional disastrous results, not to mention the caffeine-dependency introduced by non-stop cups of coffee.

Top tip for property professionals

Meetings are a vital part of how property professionals do business but they should be an aid in completing a project, one communicative tool of many, not the only way that information is shared.

If you take one example — say the creative process of design — this requires time and thought to evolve appropriate solutions; it requires sketches, exchanges of faxes perhaps. Individuals need to be able to talk things over with colleagues, to bounce ideas around within their own organisations, to access databases of information, to contact suppliers and specialists for advice.

All these things require each team member to have their own time and space in which to pursue their discipline. You may be a project manager and proud that you are able to operate from a street corner in Shepherds Bush because you have a mobile phone and a Palm Pilot. But most people like to be able to sit down at a desk at some point. That sense of place, of workspace, is fundamental. It's where you put the pictures of your kids, it's where your post arrives, where you collect back copies of professional journals and you make a determined attempt to identify that space as your own. It's no good if you're out at meetings all the time and never get to sit at your desk.

So what do all these meetings actually achieve? Client meetings should be about reporting on the progress of the job, allowing the client to ask questions. They shouldn't be formless free-for-alls where each team member tries to score points in front of the man with the chequebook. A well-executed client meeting should be boring, as there should be no new information, and it should be swift. There certainly are clients who will want to examine issues in detail, but they should be encouraged to do this at a more appropriate forum (if it's money, then a purpose led forum would be better, if it's space-planning issues, then at a briefing session with the architect); keeping 11 senior people from across a project team twiddling their thumbs while the client considers the colour swatches for the carpet is simply not an economic use of time.

Cost meetings, where the provenance of every line of the quantity surveyor's cost-plan is questioned and tested, would be much faster if everyone had managed to read the cost plan beforehand; procurement meetings where the agenda is always the same, it's just the number of the package that changes; co-ordination meetings where no-one brings any drawings; liaison meetings with no liaison; progress meetings where everyone is late ... the list is seemingly endless. You have become so accustomed to the routine of meetings that you don't use them effectively any more.

And what about recording meetings? How often do the minutes of the last meeting emerge only hours before the next one? Such minutes may satisfy the need to provide a record of discussion, but are hardly useful as a prompt for action if they're so late. Dissemination following meetings often leaves a lot to be desired too. If meetings are sandwiched together, then reporting back gets squeezed out, and others in the project team are left to carry on in the wrong direction, perhaps unaware of vital information. One suggestion is to send apologies to every other invitation to attend a meeting — you'll spend more time in the actual process of getting something done and less time doodling on the back of meaningless agendas.

Top tip for property professionals

Good effective meetings do not just happen. Everyone's role is important, whether they are running the meeting or attending it.

Effective meetings

If you are 'in the chair' here are some basic rules to follow.

Meetings as a form of communication can be used to:

- inform
- analyse and solve problems
- discuss and exchange views
- inspire and motivate
- counsel and reconcile conflict
- obtain opinion and feedback
- persuade
- train and develop
- reinforce the status quo
- instigate change in knowledge, skills or attitudes.

The key role is surely to prompt change. There's not much point in having a meeting if everything stays the same. Decisions need to be taken and this means that a meeting must be constructive. If you can develop a reputation for organising a good meeting, people actually will want to attend it. They will be keen to do so because:

- it keeps them informed and up to date
- provides a chance to be heard
- creates involvement with others

- can be a useful social gathering to allow cross-functional contact
- provides personal visibility and public relations opportunities
- can broaden experience and prompt learning.

Meetings are potentially useful. The progress of a project can, in a sense, only be certain if meetings are held regularly and are productive.

Making meetings work — like most things — requires planning. First ask yourself these questions:

- is the meeting really necessary
- should it be one of a regular series
- who should attend (and who should not).

Once you have satisfied yourself on these points, you can proceed to:

- *Setting the agenda*
  This is very important. No meeting will go well if you simply make it up as you go along. Notify the agenda in advance and give good notice of contributions required from others.

- *Timing*
  Set a start time and a finish time. Then you can judge the way it is conducted alongside the duration and even put some rough timing to individual items to be dealt with. Respect this time schedule.

- *Objective*
  Always have an objective so that you can answer the query 'why this meeting is being held?' The answer is not 'because it's been a month since the last one'.

- *Preparation*
  Read all the necessary papers, check all details and think about how you will handle both your own contribution and the stimulation and control of others.

- *Encourage others to prepare*
  This may mean instilling habits in attendees (punctuality, concise contributions from among attendees and not pausing for someone to read documentation which should have been studied beforehand).

- *People*

  Who should be there, and who should not. What roles do these individuals have?

- *Environment*

  A meeting will go more smoothly if people attending are comfortable and there are no interruptions.

Leading a meeting does not have to be done by the most senior person present. The 'chair' does not do most of the talking — in fact the reverse is true. The chairperson should be responsible for directing the meeting. Effectively conducting a meeting ensures

- the meeting focuses on its objectives
- any discussion is constructive
- a thorough review can be relied on before decisions are taken
- all sides of an argument are aired, reflected and balanced
- proceedings can be kept professional and non-contentious.

A good chairperson will see that the meeting progresses correctly, handling the discussion and acting to see objectives are met. It is essential that only one person speaks at a time and the chairperson decides who (should this be necessary).

The best meetings start well, continue and end well. Meetings should be:

- positive
- have a purpose
- create the right atmosphere
- generate interest
- be perceived as professional.

It may be necessary to prompt discussion of certain matters on the agenda. Here are some ways of doing this.

- *Overhead questions*

  These are put to the meeting as a whole and whoever picks them up starts the discussion process.

- *Direct to an individual*

  Without preliminaries — this is to get an individual reaction or check understanding.

- *Rhetorical*
  A question demanding no answer can still be a good way to make a point or prompt thinking. The chairperson could provide a response if desirable.

- *Redirected*
  This is a question asked of the chairperson straight back to the meeting: such as 'Good question, what do you all think?'

- *Developmental*
  This is a prompt to get discussion started and builds on an answer to an earlier posed question and moves it on. 'Right, Jim thinks this is too expensive. Does everyone agree, or are there other options?'

Prompting discussion between the other property professionals present is as important as control. It is the only way of making sure the meeting is well balanced and takes in all required points of view. It prevents someone returning at a later date with a remark along the lines of 'This is unacceptable. My company was not given the chance to make any representations earlier.'

Keeping control is essential. Do not, if you are chairing the meeting, get upset or emotional. Try to isolate feelings from the issue itself. Agree that it is a difficult point and concede that feelings will run high. If matters are escalating, call for silence or a short break, before moving on. You could suggest putting the problem to one side, or asking for a sub-group to be formed to deal with the tricky issue on its own. As a last resort, abandon the meeting until another time.

Top tip for property professionals

Meetings are a symptom of bad organisation. The fewer meetings the better.

Peter F Drucker (1909–) Management Consultant

At best meetings should be creative, fostering open-mindedness between the involved parties. The person in the chair should:

- actively stimulate creative thinking and ruling against instant rejection of ideas without consideration

- contribute new ideas themselves or steer the discussion into new directions
- find new ways of looking at things
- consider novel approaches and give them a chance
- aim to solve problems, not tread familiar pathways.

Some groups who meet regularly manage to do this spontaneously but others do not. For those who fall into the latter category here are some prompts:

- make it a rule that creativity becomes part of the culture of the group, or project team

- make it easy by providing feedback from the group to build in an ongoing exchange of ideas

- react to it by acknowledging what people say. If it is useful thank them, if it is negative tell them why and suggest other approaches

- give credit openly, in public, because nothing will ensure a flow of ideas more than praise and make sure it is channelled upwards as well as laterally

- make time to deal with creative ideas so that it is seen that consideration has been given to these points.

Finally, if it helps — arrange meetings as 'discussion lunches' at mid-day and provide sandwiches. This might encourage higher attendance and facilitate idea generation, problem solution and identification of opportunities.

Handling interruptions

The effectiveness of work increases according to geometric progression if there are no interruptions.

Andre Maurois (1885–1967) French writer

Be honest, if you count the number of times you leave the office with important work unfinished, wondering where the day went, the question you need to ask yourself is 'what really happened to your time?' The last chapter covered the importance of meetings and how they can swallow up hours of your valuable time. Another form of time erosion is well known to property professionals — dealing with interruptions. If you can identify what form they take you can then take steps to counteract them. Unexpected telephone calls and unsolicited visitors, invitations and requests to attend unnecessary meetings, conferences, seminars and other time consuming tasks are some of the things that can erode your time while you are at work. As a consequence your personal effectiveness is reduced and your time management is ruined.

Top tip for property professionals

When faced with impossibly busy schedules, the first and foremost tip for any property professional is to get and keep organised at all times. If you don't take control of your working day, someone else is likely to do it for you.

The more you allow yourself to be interrupted, the more interruptions you'll get. To avoid being sidetracked by interruptions, you should look at ways to avoid them or handle them effectively. If they are really unavoidable you should deal with them quickly. This could involve making a phone call, signing a document, sending an email, delegating a task to another person or simply deciding not to do something. Once you've dealt with the issue, immediately get back to the task for the day. That is sometimes easier said than done. If you have broken a particular train of thought, you will need an extra amount of self-discipline to get back on track.

Pre-empting interruptions

This is something that every busy property professional should aim to do and one way to achieve it is giving advance warning to others. By telling colleagues you don't want to be disturbed, this puts them 'on notice' that you are not prepared for unsolicited visits or calls. It might be an idea to advise those you work closely with which project you're working on. This can avoid frustration later. There is nothing worse than spending hours preparing work only to be given some additional data at a later stage which changes everything. Have you ever experienced frustration at having to re-do a job almost from scratch because you receive more information a bit too late?

One way to cut down on interruptions is to go out of your way to discourage them. Close doors — not that many people have them any more — focus on the work in hand. Telling people that you're 'in a meeting' is often all that's required. There's no need to be completely transparent and confess that you're actually in a meeting with yourself. But if this is what it takes to buy you some quiet time to get on with an important job, just do it. Most people other than the totally insensitive usually back off when told someone is in a meeting.

Refuse interruptions

Go and work elsewhere — in the car, in an unoccupied meeting room, or work from someone else's unoccupied office. Be proactive and assertive. When an interruption occurs, slot it into the priority list. If a colleague's call is related to a task at the bottom of the list (one of the ten jobs on your 'To Do' list) ask if they can be rung back later that afternoon.

Should another interruption fall into your 'Urgent and Important' bracket, you may need to be flexible enough to 'down tools' and help. Not taking action here may have implications for the rest of your team and a knock on effect for the working week. Do be prepared to say 'No'. Be pleasant but firm — there's no need to offer reasons. It is polite and helpful if you can tell colleagues when you will be available. That enables them to make a decision as to whether they can wait to speak to you, or take their query to someone else if it is time critical.

Managing interruptions

Some people just like to be liked. If this is you, you'll find it far harder to resist interruptions. If you enjoy good relationships with colleagues, staff and other property professionals, it can sometimes be at a price. You can easily get ambushed by other people because you don't like saying *No*. To help you control interruptions, work out who needs access to you and when. This applies to most property professionals because there are always many demands on your time. Issue clear guidelines as to who, when and why you can be interrupted. Otherwise let the machinery take the strain, or a colleague, but only if you are prepared to return the favour to them when they are overstretched.

If someone comes to your work station, and you are trying to avoid interruptions, ask them (politely) why they've come to see you. It could just be for a bit of idle banter. Don't encourage them to stay. If you avoid making eye contact, they will probably take the hint and wander off. Beware of getting drawn into small talk, save that for the pub or coffee break. Be ruthless with your time but gracious with your professional colleagues. If they insist on seeing you, agree to meet but say you will stop by their desk when you've finished the task in hand. That way you retain control as to when you see them and it's much easier to leave. If time runs out and the matter cannot be sorted, close down the conversation by suggesting that you'll have to continue your discussion when you've both got a bit more time.

Top tip for property professionals

The average American worker has fifty interruptions a day, of which seventy per cent have nothing to do with work.
W Edwards Deming (1900–1993) American Management Consultant

Here are some other suggestions for dealing with the four main sources of interruptions.

Telephone

The best time to call people is either before 9am or after 4pm because these are outside normal meeting hours. Set aside time each day for priority work — that means time when you don't accept calls. Field the calls by diverting them to voicemail, your mobile or someone else, if that's possible. If you switch to voice-mail leave a message advising callers the best time of day to ring so that they'll be able to speak to you personally. If the caller leaves a message for you to return their call, ring them back as soon as you can. Be courteous: ask them if this is a convenient time for them.

A time saving technique is to make your telephone calls in blocks, at the most appropriate time for you. Limit this to one or two daily telephone call periods. List the number of calls you need to make. Always have a list of the objectives of each call and include the main points you wish to raise in your notes — don't waffle. Keep a timer handy so that you remain aware of the amount of time you're spending on the phone. Whether you are speaking to the party you wish to contact or are leaving a message, be brief and clear. You should never ask someone how they are when on the phone, unless you really want to know and have the time to listen.

There is a bit of etiquette where leaving a message is concerned. You can just leave your name and a request that they call you. But if you give a brief explanation of why you've phoned and what you need to know it gives the other person time to research the answer before calling you back. It also increases the likelihood that they will ring you.

People

If you encounter colleagues whom you want to talk to, if you're pressed for time, always stand. Don't offer or accept an invitation to sit down. Carrying papers with you indicates that you have another appointment. Try replacing the 'open door' policy with a 'Visiting Hour'. With a little persuasion it may be possible to steer people towards visits by appointment only or at a certain time of day. If you are prepared to reciprocate, this can work well. Try treating others the way you would like to be treated yourself. Always check with colleagues if they are happy to accept drop-in visits.

For all of you who need to maximise the use of your working day, an effective strategy is dealing pleasantly but firmly with unsolicited visitors. When under pressure this requires tact and diplomacy. Be assertive — don't be manipulated. Asking visitors why they've come to see you puts them on the defensive. Saying 'No' to jobs that aren't yours will help you keep control of your time. Avoiding eye contact is also one way of keeping out of trouble when someone is eagerly looking for a volunteer. If their errand is not urgent, suggest fixing a meeting at a later date in their office. Implementing 'focus' time when you can work without interruption should be planned.

Assertive reactions to interruptions from professional colleagues include honesty and fairness. By acknowledging your gut reactions, feelings and wants, you also accept that others have an equal right to their feelings and needs. Aim to judge each situation on its own merits.

Requests and invitations

The key here is knowing what you need to do, rather than what you would like to do — should time permit. Ask yourself the question, could someone else attend a meeting on your behalf and get the same result? What would be the benefit to you if you accepted?

Networking is an important and valuable skill for all property professionals. If opportunities arise they should not be passed over. On the other hand if there is no discernible advantage to your attending a function, other than out of politeness or a sense of duty, then decline if your focus is on getting work done, rather than making new contacts. If it's a 'No' that you mean when offered the invitation then say 'No', clearly and directly without feeling the need to apologise, justify your actions or make excuses.

When asked for an immediate response to a query, you can ask respectfully and politely, why? If you need to think about something before you give an answer, it is perfectly acceptable to tell someone that you need time to think. You may have to consult with another colleague or superior. There is no way in a small team you can accept a meeting which will take you out of the office all day if there is no one else around that week, due to holiday or sickness. Remember to reject the request, not the person. Explain if you feel it necessary and acknowledge the feelings involved. You should watch that your body language doesn't contradict what you're saying. If undecided, say so, and ask for more time or information. Don't forget, if you cannot say 'No', what is the value of your 'Yes'?

The importance of saying 'No'

Like any good property professional, you want your work to matter, but sometimes you allow it to matter too much. If you think that by giving more than 100% of yourself to your job you are adding value, you are wrong. Of course you want to feel valued and appreciated at work and set a good example to others. But what if you rely on work for a sense of self-worth? By working harder and longer and never saying 'No', then you need to re-evaluate what you do and how long you spend doing it. If you are suffering from 'overwhelm' in the workplace, you have to begin to say 'No'. Until you do that, no amount of problem solving, motivation, time-management, or working smarter will help. As most property professionals will know, when overloaded with work, there are some things you must do, some tasks you may not want to do, and others that you simply cannot do. Whether this is because of the sheer physical impossibility of a time deadline, or because it is inappropriate for you to be doing them. Whatever the reason be honest, and say so. Assertive techniques are a great help here. Sometimes the jobs you're asked to complete aren't even yours. Practice saying 'No' to these first.

For example, the busier you are, the greater the number of telephone calls and emails you receive. Some of these calls and emails will require an urgent response. How do you develop the skills to cope with this? If you are under extreme pressure, leading up to the bidding of a new project, if you have access to a PA or someone who runs the admin for your team or department, ask for your calls and emails to be screened. Only the urgent calls and messages need be forwarded. The rest can wait until you are less pressured. You should set aside time for yourself. Agree with colleagues when it is best to do this. Sometimes entire departments agree on an hour (usually first thing in the morning) when everyone works quietly without interrupting anyone else.

Blocking out an hour or so of the day for private or 'thinking time' is both positive and helpful. Most people can get more done if there is a collective agreement than if the whole day is full of interruptions. You can use this time effectively to decide when you say *no* and when you say *yes*. It gives you space to work out realistic deadlines. If you set aside time to plan, you can deal with work more quickly and avoid overwhelm.

> **Top tips for property professionals**
>
> **Circumstances may cause interruptions and delays, but never lose sight of your goal. Prepare yourself in every way you can by increasing your knowledge and adding to your experience, so that you can make the most of opportunity when it occurs.**
>
> *Mario Andretti (1940–) Italian Racing Driver*

Dealing with professional colleagues

There is no doubt that in your professional life, working relationships do matter. In terms of personal effectiveness poor working relationships can cost companies huge amounts of time and money. Wherever your place in the hierarchy, harmonious working relationships create optimum job performance. Being a good team player does not come easily to everyone; much of what goes into being an effective colleague is learned behaviour. There are some people who have a flawless touch with others and inspire, motivate and encourage them. Some people are gifted communicators and are people oriented rather than task conscious. But for every natural there are just as many excellent communicators who got that way by attending courses and seminars and reading books on effective self-management.

Without clear communication in the workplace you can waste huge amounts of time correcting mistakes, waiting for other people's work, sitting for hours in irrelevant meetings or attempting to unravel confused lines of responsibility and authority. It leads to ineffective delegation, unnecessary checking procedures, and plunges you into tasks without planning and adequate focus. Good communicators understand people and that skill makes for easier relationships with others. With other property professionals, a successful project team leader keeps the rest of the staff motivated and focused. People can develop the capability to communicate well by mixing with others who have these abilities. Communicating so that you are understood is just the start. Being understood and understanding others isn't enough either. The essential ingredient for good communication is building relationships. This is where the snakes and ladders process links up with your networking strategy.

If you are a busy property professional and you have people working for you, you will need them to help you achieve your objectives. Without them you will fail.

Top tip for property professionals

The presence of staff is meant to reduce your workload, not increase it, and maximise the most appropriate use of your time. If managed correctly your staff will enable you to achieve far more than you could possibly do so alone.

Positive, productive relationships are easy to spot because people get on with each other, enjoy being around you and react willingly when asked to carry out favours without expecting an immediate reward. People are trusting and share information and contacts readily. Creating an environment of integrity, trust and respect requires making certain that everyone is treated fairly. Being inclusive keeps everyone up to speed regardless of whether it is inter-departmental staff or other property professionals engaged on the same project.

Managing other people

Effective management techniques are essential if you are working with other professionals on a project team. People skills are needed where the short-term aim is getting the job done correctly and on time. At a higher level there is the long-term aim of nurturing and growing junior colleagues who can lead other project teams. This goal is sometimes overlooked but is ultimately more important if you are to be a successful project manager with an increasingly buoyant work load.

Nurturing your colleagues and other property professionals is an essential management skill. If you can delegate and encourage people to take more responsibility, you will use your time more profitably and efficiently to win new business. These skills you should attempt to pass on to younger colleagues.

The main points to remember when working effectively with colleagues and other professionals are:

- create a working relationship which permits optimum job performance

- be clear — establish a goal — know your objectives
- definition of role — job descriptions help specify and protect autonomy
- maintain a flow of information — to confirm that work is progressing smoothly or raise potential problems promptly
- insist on consistency — particularly when dealing with people who tend to change their minds
- contribute to team building — it helps to have a common purpose
- show respect for others — if a person has certain traits, focus on them — prepare written work for those who are 'readers; for 'listeners' spontaneous meetings are more likely to achieve your objective.

If you look after your colleagues and staff, they will look after you. You want to work effectively and that requires support from them. Perhaps you've worked with colleagues who think by putting work above all else this makes them superior to others who do not. In fact, the reverse is true. Look at this in the context of time-management.

Ever-present colleagues and staff may have stress in their lives — damaged relationships or health issues. They believe working long days and longer nights shows they 'love' their work. Their primary motivation actually has less to do with the project and more to do with the praise and recognition they receive. They seek approval and they want to be needed.

Wanting to feel valued and appreciated at work is fine, but if you rely on work for a sense of self-worth you are exposing yourself to risk. You are at the mercy of the whims of your superiors. You desire praise from colleagues and senior staff but do not always receive it.

Introducing boundaries

When your closest friends are your professional colleagues, you are in danger of your entire support network being around work. It's inevitable if you spend all your time doing it. To discourage a presenteeism culture, where possible you should promote collegial relationships in the workplace. Friendships, like romances, should be treated with care. These can easily sour when vying with others for attention, raises and promotions.

Decisions, decisions

Decision taking is vital if you wish to be personally effective. The golden rule for busy property professionals in any managerial or leadership situation, is that the leader must be able to *take* decisions rather than just *make* them. The root of the word 'decision' means to cut. Once a decision has been *taken*, it must be *made* to work. Better the person who takes a decision, even if it is a risky or difficult one, than the one who takes no decision at all.

Top tip for property professionals

The most effective business leaders get only one out of three decisions right. But the reason they are successful is because they make even the poor decisions into good ones by working on them.

The key is action, to take the decision, and then not to waste time worrying about whether it was a bad one. Colleagues, other professionals and staff are more motivated when they have a strategy to follow. Slow decision taking is a great stress inducer because indecision saps the morale of the project team.

Difficult clients — dissatisfied colleagues

How to weed out the difficult people? Give awkward ones the choice of staying or leaving but don't risk wasting hours of time with them. You should consider ending any relationship which is consistently keeping you awake, dragging you down or draining you. Seek professional help and advice earlier rather than later. Whatever you do, deal with it.

Allow staff or project team members who want to leave to do so promptly and without regret. If a support team is to be really supportive, all members must be fully committed. Those who are not will leave eventually anyway.

Dealing with divas 13

It is the things in common that make relationships enjoyable, but it is the little differences that make them interesting.

Todd Ruthman

Successful working with property professionals who have attitude — how to achieve it? It helps if you have the skill of working out how others may react to certain situations. When faced with challenging temperaments, ideas and opinions are probably running amok. Opposing views may be strongly held. If neither party is willing to yield an inch, you could be in for some tricky exchanges.

Depending on how well colleagues know each other, it may be possible to anticipate certain situations you could be up against. It's always a wise move to have a few thoughts beforehand about what action you could take before you attend a meeting — certainly before you open your mouth. The smooth and smart operator keeps a list of possible motivational actions ready to implement if you have to pour oil over fractious colleagues. Depending on their mood, the degree of success you are likely to achieve will vary.

Top tips for property professionals

When dealing with divas be prepared to be flexible. Don't always use the same methods and tactics, particularly if they are allowing their behaviour to let them down.

An intransigent attitude from a member of the project team could be defeated by a flexible approach from another person. It might disarm them and bring about a swift conclusion to the argument. If progress on the project is the objective, then whatever method employed to bring this about is the right one.

Communicating with people in general — and especially when coping with challenging individuals — requires tact and diplomacy. Evaluating what works best for you and suits your way of operating is just as important as knowing how to work effectively within your project group and professional discipline. Inanimate objects can be dealt with efficiently and in reasonably quick time — but when dealing with people (and the more talented they are the more complex the individual often is) matters simply cannot be rushed.

Top tips for property professionals

Business is not just doing deals; business is having great products, doing great engineering, and providing tremendous service to customers. Finally, business is a cobweb of human relationships.
H Ross Perot (1930–) American Businessman and Politician

Time and tide

Working to achieve the best from professional relationships is difficult. There is a material element to all dealings - with colleagues, staff, clients and professional advisors. This is why it is so challenging. There is a limit to how much time can be spent pursuing a trouble-shooting strategy. Someone, usually the client, will have the ultimate sanction. 'Cancel the project', 'dismiss that team', or 'stop dealing with that contractor'. Until you receive those instructions, most likely you will try your best to pour oil on troubled waters and strive for a satisfactory solution.

Your approach to dealing with awkward characters in the property and construction industry requires skill at working on 'special' relationships. You may all be from different professions with varying expertise, experience and qualifications; it's a bit like getting cats into a sack. There is bound to be some scratching and clawing. However there is a limit to how much time and money can be spent on making things run smoothly among awkward personalities.

When colleagues among the project team cause pain and anguish to others, this is going to cost hours of productive work time. It will upset the client and may well have a knock on effect, if some of the equipment and processes are being badly administered by these unhelpful characters. That is why it is essential you are able to deal with conflict situations effectively and bring about a sensible resolution. There are a number of things you can do to reduce the nuclear fall out that bad relationships on projects can cause.

To deal effectively with a disparate group of people, it helps to be able to recognise certain types. A *High Maintenance Person* is someone who, when a one-off gesture is made, however small or simple they take it as an ongoing commitment that this act will be carried on indefinitely. They assume you are delighted to have been of service. If you've been caught unawares by one of these people, it will take you by surprise — but only the once. You will learn quickly not to make similar gestures since they seem to operate under the impression that you have few other demands on your time.

High maintenance people are easily spotted once you've got their measure. Colleagues, clients, staff and others can be the most difficult people on earth, if they have high maintenance people tendencies. You need to be able to identify them and warn others against falling into similar traps. One way to keep high maintenance people in check is to use assertive communication techniques. These are essential when dealing with troublesome people. You can save yourself many hours of time and avoid lots of hassle if you develop the ability to take swift control of an otherwise unmanageable situation.

Top tips for property professionals

You don't develop courage by being happy in your relationships everyday. You develop it by surviving difficult times and challenging adversity.

Barbara de Angelis

Getting a grip

What does assertive behaviour help you to do? It enables you to get into the habit of expressing ideas positively and communicate them in a professional way. Assertive communication helps you to satisfy the needs and wants of the other parties involved, assuming they are calm enough to listen to reason. If you are assertive, you will show that you are able to stand up for yourself without violating the other person. You are most likely to do this, even in the most awkward of situations, by making positive statements in an honest and direct way.

Statements such as:

- What I'd suggest we do first is ...
- There is a way I can probably help you to ...
- I'll check for you that ...
- It will help to get to grips with this straight away ...
- How does that sound to you ...

show the other side that you are positive, confident and prepared to be proactive on their behalf. It will go a long way towards getting them to calm down or move towards your viewpoint.

Often the aim of your colleagues' aggressive behaviour is quite the opposite. Bad conduct can frequently be the desire to prove their superiority, to threaten, or to defend against what they perceive as a threat to them. These people express their thoughts, feelings and beliefs in unsuitable and inappropriate ways. Even if they are in the right, their openly hostile manner drives potential allies away.

Assertiveness can be interpreted as:

- appropriate behaviour
- self-confidence and self-esteem
- communication
- being positive and proactive.

When you behave assertively, you leave others in no doubt that you are in control, but not controlling. The key to being assertive is that, in any difficult situation, you leave the exchange feeling good about yourself and the other person involved (I'm OK, he's OK). The aim is for a win–win outcome in terms of mutual respect and self-respect.

Communicating in an assertive way, you say what you mean and you mean what you say by giving clear, straightforward messages. You will show that you can:

- be direct
- be appropriate
- take responsibility

- remain calm and in control
- be willing to listen.

If you can practice getting your assertive technique right, a lot of anxiety you feel about facing challenging situations will disappear. Your confidence will grow and you will no longer be timid in potentially confrontational exchanges. Your feelings of guilt, embarrassment and frustration will evaporate.

Top tip for property professionals

Assertiveness means standing up for yourself and your rights, without violating the rights of others. It allows you to understand yourself and be who you really are.

Distinguishing characteristics

The contrast between aggressiveness and assertiveness is marked.

An *aggressive* response is a put down. It is often a personal attack, tinged with sarcasm and arrogance.

An *assertive* response is a reasonable objection delivered in a polite but positive manner.

When faced with some difficult situations, an assertive response sometimes requires that you postpone tackling the issue. Why does this help? It shows that you have courage and confidence and are not being railroaded into action without appropriate consideration. It can help you:

- create a boundary
- buy yourself some time
- avoid rambling excuses if you don't have the facts.

You are being assertive if you refuse to be pushed into a situation you are not ready to tackle. A confrontational member of a team who is revelling in creating a conflict situation should not be given satisfaction. Just because he has allowed his 'inner child' to have a full-blown tantrum in the office, you should remain uninvolved. Resist the temptation to shout back 'Oh why don't you just grow up'. This won't help because you will be behaving in an equally juvenile manner.

Summoning all your confidence and skill, a smooth operator like you will be able to buy yourself some time. Show no signs of annoyance, sarcasm or arrogance, but allow your voice to register concern and interest for your difficult colleague. What you are attempting to do here is to deal with a badly behaved child in an adult manner. With a bit of luck and breathing space, your colleague may come to his senses. If he does behave more sensibly, you can then begin to deal with the situation (and him) in an adult way.

What is being described here is the theory of *Transactional Analysis* (sometimes known as TA). It was first developed by Eric Berne, an American psychiatrist, who through observations of his patients formed them into a psychological model. It clearly showed certain patterns of thought and behaviour. His theory suggests that whatever our age or status, there are three ego states from which people operate. These are *parent–adult–child*.

1   *Parent*
    The mixture of behavioural codes that you were brought up with during your formative years — the *taught* element.

2   *Child*
    This represents how you received those rules which surrounded you, the feelings they evoked and how you *felt* about them.

3   *Adult*
    The result of the influences on you, your experience of life, your perceptions and realisations — the *thought* element.

The *parent/teacher* messages are those that usually contain the 'don't', 'do', 'mind out' and 'watch it' words. This type of expression tends to lay down rules and give little or no explanation. They are often critical too so the person feels little and undermined. If you are dealing with someone and giving parental messages, you will be telling them they should control themselves, admonishing them and advising them that they should not question authority. Perhaps you could go so far as to tell them they are a nuisance.

On the receiving end of this, if you are trying to respect the authority figure, you will be telling yourself that you 'should' and 'must' improve, do something better or quicker, finish your work on time, show that you are cleverer than your colleagues. A lot of this conveys negative messages and does not include praise. This is why childish behaviour at work can be the result of reprimands, causing the employee to feel bad about themselves and gets in the way of them being assertive or positive.

When you communicate with others in the parent ego state, you are most likely to be criticising them. You probably won't be offering constructive feedback and will be coming across to them as aggressive. 'You shouldn't ...', 'you've no right ...' 'you're always ....' If these are the sorts of remarks you are making to an angry colleague, this is not going to help the healing process in the slightest.

When professional colleagues give out parental messages to others, it causes feelings of humiliation and resentment, and sometimes outright rebelliousness. Communicating in the parent ego state, your remarks will most likely sound

- dominating   (you simply cannot ...)
- inflexible   (you have no choice but ...)
- patronising   (when I was in your position ...)
- intimidating (you'd better sort it out ...)
- commanding   (you will do it ...).

This means you are not being assertive, but aggressive, and the person with whom you are dealing is prevented from being assertive too.

Now, look at things from the *child* ego state. This is determined by how you felt about things when you were very young. What emotions did you feel: happy, sad, frightened, excited, curious, frustrated, confused, angry, loving, rebellious. You will react to the 'parent' remarks with one of these emotions. Children are freely expressive creatures and this is what is not expected in the adult world, particularly in the workplace where a degree of masked emotions is required.

To be assertive you need to be emotionally and physically expressive because this shows a positive attitude in life. Life is an opportunity, not a threat. It should be enjoyed and in most cases happy experiences outweigh the sad ones.

The project team co-worker who is behaving in a childish way is giving way to anger, jealousy and frustration by hitting out at colleagues, superiors or staff. Their behaviour, while spontaneous, is totally selfish. They are showing no regard for the feelings of anyone else. When communicating in a childish way, work colleagues will sound as if they are:

- moaning   (why me? it's not fair ...)
- demanding   (I want ...)
- dealing   (I will if you will ...)
- angry   (storming out of rooms, slamming doors ...)
- scowling   (it's not my fault ...).

When someone in a professional situation reacts in a totally disproportionate manner to what is actually happening, they are behaving in their *child* state. Something sets off that reaction in them: threats, humiliation, frustration or a sense of unfairness. Hence the expression 'Oh he's just thrown his rattle out of his pram'.

The adult ego state is the most desirable one and the best one from which you can behave assertively. You probably operate in it most of the time. It is the rational approach — combined from the thought and learned experiences which have made up your character and personality. It governs how you live, and how you relate to and accept others. Your adult state is non-judgmental; it takes account of your own failings and allows other people's rights. This ego state relies on knowledge and learned behaviour and when operating in this state people behave objectively.

People conducting negotiations successfully always operate in the adult ego state. This is why they are able to be assertive, but above all, reasonable. Their remarks may include some of the following:

- understanding   (I see your argument ...)
- solutionist   (what would you do ...)
- revealing   (it is disappointing to learn ...)
- defining   (so this is how it happened ...).

Strengthening your ability to deal with people from the *adult* ego state, will enable you to develop your assertiveness skills. There is no need to completely suppress the other ego states — as long as you use them wisely. The parent state should be used to show dangers and opportunities. The child state will reflect your experiences throughout life: your emotional responses.

Being aware of the three ego states should help you to acknowledge your own behaviour and that of others. If you can work out which ego state someone is in, it will help you decide why it is, or is not, appropriate for the situation you are currently dealing with. It will show you why you behave towards some people in a certain way. Assertiveness is having respect for yourself and others.

> **Top tips for property professionals**
>
> The best index to a person's character is (a) how he treats people who can't do him any good, and (b) how he treats people who can't fight back.
>
> *Abigail van Buren American Advice Columnist*

When faced with an angry person, you will need to behave assertively, not defensively. If they raise their voice, lower yours. If they are moving about, using wild gestures, be still. Use open questions when enquiring about their problem. Look for ways to resolve the issue by offering some possible solutions. If a cooling off period is required to allow a reduction in temperature, suspend reaction and agree a time to resume the exchange.

It's a sensible idea to allow your colleagues to express their anger if they need to 'get something off their chest'. In many cases this could be all that they wanted to do. When they have finished ranting and it's your turn to respond, do so in a calm manner. Show them that you have listened by repeating back to them what you took to be their main points of grievance. Tell them that you are prepared to do something to help.

By remaining relaxed, you can diffuse many awkward situations. You could outline the problem by summarising what the other person has said to make sure you've understood it correctly. It is important to discover the root cause of the issue. You will need to ask one or two questions to find out why, when, where and what has caused the particular problem. For instance, was it the action or behaviour of another person?

> **Top tips for property professionals**
>
> Characters do not change. Opinions alter but characters are only developed.
>
> *Benjamin Disraeli (1804–1881) British Prime Minister*

New ideas and approaches

When dealing with divas, check that your behaviour is a strong, positive influence. What if one of your team suggested tackling a situation in a particular way, and your reaction was one of annoyance. Don't respond with 'Oh really, what a ridiculous idea, that'll never work'. At a stroke you have crushed a potential ally, caused the person humiliation and probably forced them to behave aggressively towards you at the next exchange.

New ideas are delicate things and require a response that is from the adult ego state. They can be dealt a terminal blow by an ill-judged remark, or because of anxiety they may never be aired, or even see the light of day. Have you ever decided to bite your tongue and say nothing because of fear of rejection and humiliation? Would your idea perhaps have saved a situation from getting worse, or brought about helpful changes in your organisation?

New approaches and ideas are the fuel that projects need to keep them up to date and delivered on time and on budget. Never be afraid to air them. If you have a wise project manager he will encourage innovative ideas from his team. After all you are all united in wanting successful delivery for the client.

Not every new idea will be welcomed with open arms. Don't worry about this and don't take it personally. Resist falling back into the child ego state. Many people fear change and it is often the case that new things don't always work out ideally. With positive encouragement and open minds, discussion can sometimes lead to an unacceptable idea being adapted and then adopted. It could even bring about a better solution than had first been anticipated.

So if you have a bright suggestion to make which could solve a potentially confrontational situation, think about it first, before opening your mouth. It may be that you simply wait a few moments before speaking out at a meeting. If you think it is a particularly strong idea, test it out on a trusted colleague first. If they end up rolling around on the floor convulsed in laughter, wait a while. Maybe your idea needs a bit more development.

It is always sensible to consider not only the short term but also the long-term implications of the solution. It might for example rectify one aspect of the dispute but cause much wider ranging effects in other departments. So do give it real consideration first.

If you decide it is worth putting forward as a potential answer, make the suggestion with conviction and confidence. If it is not taken up, don't worry. It might not have been the right time, or you may not yet have reached the point where you are recognised as a 'new ideas' person.

Saying 'may I make a suggestion' politely and appropriately is usually viewed as helpful. Your fellow team members will be grateful that you have given thinking time to their predicament.

If you practice putting forward well considered, well reasoned and researched ideas for potential solutions, you will soon gain a reputation as being one of the 'lateral thinkers' in your group. Creativity is what everyone wants from construction and property professionals — that is what will ensure progress in your career.

Anger, frustration and other behaviour

Behaviour is the mirror in which everyone shows their image.
Johann Wolfgang Goethe (1749–1832) German poet

It's a very good rule to bear in mind whatever your profession, and particularly with busy people, never personalise a difference of opinion. You will not get a fair hearing if you do. Do not react instinctively but listen to the content carefully. When responding, focus on what was said, not on the way the person behaved. It is important to respond in an impersonal way when you are dealing with aggrieved people. Where it is necessary to disagree with a colleague, they are more likely to accept neutral comments. Always thank them for their contribution. If it becomes necessary you may feel it appropriate to point out politely that their idea may not work. Explain that 'under the circumstances', 'due to the amount of time we have available' or some other reason, that you will have to consider other suggestions. The discussion can then be continued without rancour.

Combating desk rage

There are probably very few of you who, if you are being really truthful, can say that you've never allowed the frustrations and irritations of work to make you feel stressed and horrible. The occasional office tantrum is to be expected but if there are signs that pressures at work and life generally are sending people into rages at an increased rate, you will find it affects the atmosphere in the workplace.

There is evidence that bullying at work is on the increase, and conflicts among colleagues are rising. When you consider the increasingly busy, complex lives people lead, this situation is not entirely surprising.

But why has it reached such epic proportions? Is there anything you can do to remedy the situation? Organisations competing against each other for a greater share of the global market place are constantly operating in a changing environment and this also applies to the property profession. The rules about what you are expected to do and how you are to do it are altered time and again. Technology may have made your life easier but it piles on the pressure for you to deliver things much faster and right first time.

Social changes also add pressure. Most couples work and when combining job responsibilities with the added stress of caring for children or the elderly, no wonder you often feel burned out and in an almost permanent state of suppressed fury. There are added pressures to living and working in multi-cultural environments. Different behavioural patterns and customs often add stress where it did not exist before. With the rise of the compensation culture, and people's added awareness of their rights, your expectations are also rising. If you happen to work in a client facing environment, as many property professionals do, you are only too aware of how angry people get when their expectations are not met. The result? Something's gotta give. A pressure valve lets off steam and someone let's rip with their temper.

Work does of course have enormous potential for winding anyone up. Why does your computer crash an hour before an urgent presentation? Does it seem that your phones ring more often when you have to concentrate hard to understand a difficult report? It is often just little things that tip you over the edge and into inappropriate office rage. It might be people interrupting you when you are trying to work, or a colleague eating something noisily sitting next to you. Although the rage may be relatively short-lived, the bad vibes continue causing long term damage. If you are unfortunate enough to work with a boss who loses his 'rag' regularly, colleagues become anxious, demoralised and defensive. A tense atmosphere is the overriding result.

Three main factors contribute to workplace anger. These are:

* personality clashes
* varying work styles
* differing goals.

What you need to do to address these issues are policies which:

1. Protect staff from abuse from bullying colleagues or when dealing with irate customers

2. Have procedures which deal with the aftermath of aggressive episodes, including de-briefing or counselling

3. Have guidelines which set out what is and what is not acceptable behaviour in the workplace, ie no swearing

4. Hold regular 'surgeries' to find out what issues staff are seething about

5. Communicate fully and frequently about proposed changes in the workplace, to combat rumours, gossip and allay fears.

Anger has such a negative impact on people too. It lowers their morale, causes dissention and you end up feeling like you're walking in a mine field all the time. When you are new to an organisation, you generally look around you to see what other people do in terms of what is acceptable and what is not. Once you understand how things are done, you relax and being to settle into an appropriate working pattern. It is the responsibility of those in management positions to make it clear that aggressive behaviour is not tolerated. If you don't know what is and what is not allowed, how can you work out what the rules of engagement are? You certainly won't know what you are expected to put up with from colleagues and co-workers.

Where there is ambiguity, people get stressed because they have to deal with situations which they may not really have to endure. One suggestion that might work towards helping people remain calm, is to keep a 'rage gauge'. This is a record of how angry people really become in certain situations. Group meetings are held where colleagues are encouraged to share their feelings in a healthy and positive way. It is only by communicating effectively with others that you can attempt to solve the issues that get in the way of your ability to work harmoniously with your colleagues.

De-stress measures

Anger makes behaving badly very easy to do. If you think you're about to lose control one way of regaining some composure is to look around you at the things you can control, such as your personal environment. Cleaning, filing and organising is very therapeutic. A serious bout of clutter-busting is beneficial to the temperament. Once you've cleared out a whole load of old paperwork and shoved it through the shredder, you will begin to calm down. Vigorous clearing and sorting can be soothing. When everything around you is in order, your mind often feels correspondingly calm. Have a bin-blast: clearing, cleaning and chucking out is a way of clearing out the confusion and anger and you will feel considerably better.

If you are working long hours at a desk, make a decision to get up and stretch regularly and take some deep breaths. Taking regular exercise helps. Even climbing the stairs rather than taking the lift is good. It improves your frame of mind. Eating sensibly and frequently is good. Whether stress is mental or physical, good exercise and eating habits prevent the body from increased stress levels. Get enough sleep. Lack of sleep clouds the judgment and deprives you of the ability to think clearly and make decisions. Don't forget laughter. Physical exercise is good but laughter is great therapy and one of the best tension releases there is. Regular laughter can permanently lower the heart rate and blood pressure. Try starting the day with a smile. If you smile at people, you will find it is amazing how addictive it is. People respond quite naturally to a happy face and it makes you (and them) feel better.

Top tip for property professionals

People don't change their behaviour unless it makes a difference for them to do so.

Fran Tarkenton — American Football Player

Take control

Research shows that the less control you have in your life, the more stress you feel. Certain jobs are high in psychological demand (mental challenge) and low in decision making (control). These are the most stressful. If you work in such an environment your stress levels are going to be far higher than someone who is more in control of their destiny. What do you have to cope with in your working day? Is your role largely autonomous and pro-active? Or does it rely on others delegating work to you which you have to complete in a certain time? This will affect the amount of stress you feel at work. When pressure becomes intolerable, it can be incredibly difficult to see a way through it. The best way to cope is to stop, take a step back and take each task one at a time.

Coping with moody colleagues is not easy. You take it out on family members, and can be short-tempered with those who do not deserve it. You probably feel tired all the time, and when staff are apt to burst into tears or fly into rages for what seems like no apparent reason, it can make you feel like doing the same. Harbouring feelings of anger, aggression, conflict or frustration without any particular excuse is a typical example. Venting feelings of frustration occasionally is not harmful, but beyond a certain limit it has a marked effect on everyone with whom you work, mentally as well as physically.

Learning how to react to stressful situations that cause anxiety and anger is vital. Acting positively, assertively and constructively will enable you to stay in control and reduce other people's behaviour to more acceptable levels. Remaining relaxed and calm will help them regain composure. The body can't be stressed and relaxed at the same time. If you are feeling overloaded, and can't take on any more, you must learn to say No. Prioritising your tasks is important — delegate something if possible. Draw up an action plan, and set realistic goals. This is dealt with elsewhere, as is the necessity of learning to say 'No'.

If you are trying to help a colleague deal with their behavioural issues, suggest that they keep a diary. Making a note of how you feel on a daily basis — happy, unhappy, unsettled, in control, out of control, anxious, frustrated, confident, successful will help to gauge their moods and give some warning of whether an outburst is likely. Help them to analyse the results. When did they feel most happy and on top of things? What had they done that day? What factors had influenced them? How much time did they have to themselves? Had they taken control of a situation which had had a positive outcome? Who had they spent their working time with?

Top tip for property professionals

Our fatigue is often caused not by work, but by worry, frustration and resentment.
Dale Carnegie (1888–1955) American Author and Trainer

Carry out this exercise — answer yes or no to the following situation. Where you have answered 'yes', score them between 0–10 points, depending on the stress rating you give them.

1   Have you ever been landed with a huge piece of work, just a week before going on holiday?   Yes/No

2   Is your company being forced to make redundancies?   Yes/No

3   Do you encounter unreasonable or unsympathetic behaviour from colleagues/family/spouse?   Yes/No

4   Is your boss refusing to support your bid for promotion?   Yes/No

5   Do you regularly have to work evenings and weekends?   Yes/No

6   Are you made to feel awkward if you request time off for family reasons — eg elderly parents, spouse or children unwell?   Yes/No

7   Have you had to miss out on significant family events in the last 12 months due to work overload? Birthdays, speech days, wedding anniversaries?   Yes/No

8   In the last year have you had to cancel a theatre or concert/holiday/weekend away due to pressure of work?   Yes/No

9   Do you regularly have sleepless nights because of work?   Yes/No

10 Less decisive, unreliable memory, loss of concentration?   Yes/No

How many of these issues could relate to you? Would you have scored highly in the 'temper chart'?

(1) 0–40   OK
(2) 40–70   Be careful
(3) Over 70   !!!!!!!

Suggestions for anger management and mood control

1   Take some time off. It doesn't matter if it's a weekend, or just a day. Distancing yourself from the cause of the trouble (other difficult colleagues) is often a great help.

2   Walk to work. People who walk regularly not only have stronger legs but a larger number of brain cells.

3   Leisure time — how much of your working week is spent doing nothing? If the answer is less than a day, add another four hours to the time schedule.

If work is the dominant feature of your life, or it seems to be heading that way, you should keep an eye on the situation. There are times when you simply must work long hours at the office, or you have to stay away from home on business. During those times you have less self and home time. You automatically become more tense and you need to make sure that this does not become a pattern. If the situation is temporary, you can use your common sense to compensate in other ways and keep your temper under control.

Some companies now monitor the working hours of their professionals. The process is designed to identify people who work over-long hours. If the professionals seem unable to adjust their workloads themselves, after a period of time they are offered counselling on how to work more effectively.

Top tip for property professionals

We can learn to condition our minds, bodies and emotions to link pain or pleasure to whatever we choose. By changing what we link pain and pleasure to, we will instantly change our behaviour.
Anthony Robbins (1960–) Author and Performance Expert

Being assertive is a way of bringing situations under control which could otherwise get out of hand and cause unacceptable behavioural outbursts. Assertive behaviour is not the same as being aggressive. But it is the opposite of being passive. If you are prepared to speak up and let people know where you stand it may be an effective way of diffusing a potentially difficult situation. It may be that others have simply not realised the situation from your own perspective. If you don't tell people how you feel, how can they possibly know? It isn't that easy to be inside another person's head. The other person may not be as difficult as you imagine. By informing people of your wishes, aims or desires, it completes the picture. There are two sides to every situation and disclosure is the only way to sort a situation out.

Separate fact from opinion. In order to bring about a solution or compromise, it is essential to have the relevant facts. Hearsay, gossip, personal views do not help the negotiating process and can be viewed by the other side as irrelevant. Recognise things can be different for other people. How often have you reacted to a situation because you think you are in the right? Ask two people to describe the same incident. Don't forget your perception of a situation can be quite unlike another's version. Be prepared to negotiate. Most successful defusing strategies are achieved through effective compromise. If you're prepared to give something to ensure a win–win situation, you may well be agreeably surprised at the outcome.

Taking a break

When you are overworked, and you feel your temper is permanently frayed, you tend to think longingly of your holidays. But often holidays are ruined because you can't say 'No' to your work. Many busy people find it difficult, if not impossible, to take a real break. You are incapable of turning off from the stimulus that comes from your work. Although you are on holiday, you clutch your laptop as you board the plane, and while you are away you check your emails daily, retrieve messages from your mobiles and talk to staff, colleagues and superiors at the office. This is no holiday. It's a continuation of your working life carried on from another location. There's no rest or benefit going on here. By succumbing to the daily information glut emanating from your office, you fail miserably in taking that well-earned rest and are not fully participating in the longed-for holiday.

It is vital to cut back drastically on the amount of communication with the office while away. It may not be possible to go completely 'cold turkey' but at least set some realistic boundaries. Make it clear to colleagues and associates that while they can remain in contact,

---

Top tip for property professionals

With me a change of trouble is as good as a vacation.
David Lloyd George (1863–1945) British Prime Minister

---

communication should be kept to an absolute minimum. Another important point to consider: why is this vacation important? What do you need from the trip? If you're on holiday with your family, what do they need from this trip? Are you going along to be a fully participating member of the group? Remember, the best gift you can give anyone is the gift of your time and yourself. Because you are prepared to work hard with long hours at the office, when you are on vacation quality time with families and friends is just as important. If you block out sufficient time, at regular intervals, and schedule it to share with those closest to you, this can have a real impact. It makes for good, strong relationships from which everyone benefits.

A lot of people use holidays to catch up on reading. But you should confine it to what you want to read. While some people pack too many pairs of shoes and loads of clothes, you should include a number of books that you've had on the 'holiday reading' list for a while.

An ideal holiday policy is to blank out completely. If you're gone, you're gone. No email, no phone, no work related updates of any kind. For this to work really well, the holiday should be two weeks' long. For most near-workaholics, the first few days are taken up unwinding completely. Your mind is still in office mode. After that you begin to relax. By the second week you're on a different planet. When it's time to come home, you'll be almost excited at the prospect of returning to work. Anyone who has lots of responsibility needs to get away from the workplace for a period of time. It is a rejuvenating experience. More important, you're probably not the only ones to benefit; your colleagues and staff deserve a break from you too.

Top tip for property professionals

A vacation is over when you begin to yearn for your work.
Morris Fishbein (1889–1976) American Physician

It might be appropriate here to draw the distinction between trips and travelling. For the majority of you, taking a holiday means going on a trip. You know where you are going, you've chosen the location, the amenities and surroundings. To be a success, it meets your expectations. If it falls short, it's a disappointment. Travel, on the other hand, is not about such superficialities. Travelling is about seeing beneath the surfaces. To quote Paul Theroux, the novelist and travel writer, 'travelling is about self-discovery. It is personal, literary, and mystical. It's about the challenge of a self-led adventure in an unfamiliar landscape. It's sometimes uncomfortable and often upsetting. But it's worth it. If you make a meal of your adventure, your spirits will be revived. You'll walk away from the experience with an insight into yourself, or into the world.' In reality, cramming all that into a two-week holiday is an impossibility. But if the opportunity were to present itself — say a three month sabbatical, or a period of garden-leave — wouldn't that be the perfect way to look at life from a different perspective?

Creating personal reserves

Another fairly simple way of dealing with difficult colleagues or work place pressures is to try distancing yourself from stress inducing situations by creating personal reserves. Pacing yourself is one way, if you have a number of difficult tasks ahead. Concentrate on one or two things at a time. Letting go of a situation is not failure. It is a positive and powerful course of action. In order to renew your energy and focus your attention, usually something has to be given up.

Do something relaxing. Creative ideas often come to you when you are away from the work situation. The solution to a problem can appear if you give yourself time to allow your subconscious mind to operate.

Part 4

Managing Others in the Property Profession

Working with others

Smile, for everyone lacks self-confidence and more than any other one thing a smile reassures them.

Andre Manois (1885–1967) French Writer

Confidence and assertiveness when dealing with others

Whatever the situation, one of the most important skills for property professionals to acquire is confidence and assertiveness in dealing with different types of people. Whether they are colleagues, superiors, clients, subcontractors or associates, the ability to conduct yourself appropriately with other professionals is highly desirable.

This includes creating the right impression, building confidence and self-esteem so that you can manage relationships with high maintenance and potentially difficult people more easily and successfully.

Top tip for property professionals

Remember — you never have a second chance to make a first impression.

First impressions

Within a few moments of meeting, other people make assumptions and judgments about you. However hard you may try to avoid doing so, you are likely to make an instant decision about someone you meet because of the way they look, speak or what they wear. This can impact hugely on a working relationship where professionals are involved.

Research says that when making an entrance:

* 55% of the impression made is how you look — posture and what you wear
* 38% is the energy and enthusiasm — body language, tone of voice
* only 7% is what you actually say to a person.

Visual impressions are more important than oral messages

You are working in a professional environment, so if you sense there are likely to be personality issues, it is helpful to contain any potentially difficult situations before they get out of hand. It is an enormous advantage if you can remain calm and in control, so that even if others are being provoking, you can take steps to limit the trouble they may cause.

Getting off to a positive start makes things easier. A good beginning not only affects the quality of the encounter, it affects your confidence too. Confidence requires preparation and needs to be actively worked at to ensure you achieve the right impact. If you are well prepared, mentally and physically, for an important team meeting, you will appear more confident, calmer, and better able to handle any difficult situations that may follow.

This is not a question of tricks or gimmicks. It's about being business-like and professional and aware of the importance of everything going well in the early stages. Having the intention is the first step towards achieving it.

Top tip for property professionals

Clothes and manners do not make the man; but when he is made, they greatly improve his appearance.
Henry Ward-Beecher (1813–1887) American Preacher

Looking the part

If you want to be seen as confident and self-assured among other property professionals, capable of diffusing difficult business situations in a cool professional manner, using open body language will make you more persuasive.

Stand upright, balanced on both feet with your weight evenly distributed. Remember, your body is an instrument — it conveys every emotion. A good tip is mirroring gestures. These are great for creating a good first impression with a challenging person. By copying what the other person does, it sends a positive message, endorsing what information they are conveying.

If you encounter someone being difficult and making aggressive gestures, such as pointing, shaking their fist, don't employ mirroring gestures here, it could result in the situation escalating. Actions speak louder than words and body language speaks volumes when faced with awkward situations.

If you want to create a favourable impression with someone, your body will quite naturally point towards them — your face, hands, arms, feet and legs. These gestures can be subconscious, but the other person picks them up quite easily.

Have a good look next time you're in a group of people — whether a social or work situation. Observe how individuals position themselves when communicating with each other. They naturally angle themselves towards the person with whom they are trying to create a positive impression, and turn away from those who they are seeking to avoid.

Eye contact

Making the correct sort of eye contact in work situations is important. You are probably dealing with someone you don't know very well so there are a number of things to remember. It is natural to look at people from eye to eye and across the top of the nose. This is the safe area to which eye contact should be confined. With friends, away from work, this area of vision increases to include both eyes but also downwards to the nose and mouth. If you're nervous, you should avoid staring at someone when they're speaking. On the other hand, looking away completely, blinking or closing the eyes for longer periods than normal, can indicate shyness — an attempt to block the situation.

In a meeting when conveying information, pointing to something you are discussing (such as a model or drawing) directs attention towards it, and away from yourself, if you need a break from all eyes being focused on you. You can then bring the focus back again by lifting your head and engaging eye contact again. This is helpful if you need to change the emphasis of your meeting.

Keep your head level and your gaze open in difficult business encounters. Holding your head straight both horizontally and vertically gives the impression of authority. A friendly gesture when listening to someone addressing you is to tilt your head slightly to one side. This gives the message that you are attentive and open to what is being said.

Top tip for property professionals

One important point to remember, you've been given two ears and only one mouth — so use them in that proportion.

Bear in mind that if you spend twice as much time listening as talking, you will create a positive impression. Other people will regard you as a skilled communicator who can operate effectively in potentially awkward situations.

Clothes matter — aim for well groomed, rather than high fashion.

Following high fashion trends is not appropriate in a corporate setting. It's far more helpful to have well kept hands, clean and tidy hair and neat clothes.

Give a smile — it's free

Many people have the most wonderful natural smiles, but due to nervousness or apprehension, they don't let them show. When greeted by a smiling face, people notice. More often than not the natural reaction is to smile back. This gives a positive impression in challenging situations. Why not come across as being pleasant, attractive, sincere and confident? It may reduce the fear and anxiety of those with whom you are making contact.

Good manners

Good manners never come amiss. How do you feel if someone bothers to say thank you if you've travelled some distance to see them, or made an effort to attend a meeting when you are very busy? When dealing with other professionals who also have hectic work schedules, a business-like approach will stand you in good stead.

Turn up on time when you have an appointment. The ability to be punctual at a first meeting gives the overriding impression that you are well organised and capable of delivering (even if it is just yourself). Turn up late, however, and all that will be remembered is that you missed the appointment. It may sound harsh, but this could jeopardise a future working relationship. A lot of hard work will have to be done to help redress this, so don't make things difficult by an overly casual approach.

However organised you are, you should always allow extra time when travelling, to avoid stress. If you arrive for a meeting in a fluster and out of breath, you'll be in the wrong frame of mind and won't be in control. Appearing cool, calm and collected is well worth the extra investment of getting out of bed an hour earlier.

Top tip for property professionals

Men in general judge more from appearances than from reality. All men have eyes, but few have the gift of penetration.
Niccolo Machiavelli (1469–19527) Italian Author and Statesman

---

Pay attention

This may sound like unnecessary advice, but it is surprising how many people can't stop their eyes straying when someone walks past an office or a commotion takes place outside. Keeping your eyes and ears directed towards the client, or whoever is speaking, is vital if you are in an important meeting where temperaments are clashing. Don't relax, show that you are giving the situation the attention it deserves. By showing that you are concentrating, this will create confidence in your ability to cope. This is particularly important in meetings where a number of characters are competing for the upper hand.

Mobile phones

Always switch off your mobile phone. There is no better way to irritate people in a business meeting than being interrupted by an unwanted bleeping coming from your pocket or bag. Despite constant reminders, people still forget to switch off their mobiles and it never fails to annoy others. Don't compound the sin by answering your phone — that's lethal. This rule applies the other way round too. If one of your colleagues or contacts has the insensitivity to receive calls and messages throughout a meeting, it's an insult. It shows a lack of respect not only for the occasion but also for the others present and creates completely the wrong impression.

There are occasions when such interruptions are unavoidable, for instance if you are awaiting the final details of a report, or the outcome of an enquiry. If you have to conduct a meeting while expecting an important call, have the courtesy to say so and announce the reason why you are awaiting some information. Then, when the phone call does interrupt the dialogue, there is no need for embarrassment. Just excuse yourself for a moment while you take the call, make a discreet exit and be brief.

Top tip for property professionals

I was always looking outside myself for strength and confidence but it comes from within. It is there all the time.

Anna Freud

Prepare for the occasion

If you are keen to establish good relationships with other property professionals, some of who can be highly competitive, you should have a plan for boosting your morale. With healthy self-esteem you will have the confidence required to work closely with challenging individuals in the most favourable and positive way.

Building yourself up so that you believe in your ability to succeed is very important. Behave and look as if you have already achieved your goals and you are half-way there. Confidence breeds confidence and as it develops, it will become natural to you and have a positive impact on others.

It doesn't matter who you are, people make judgments based on their first impressions. One of the key reasons why you should spend time and effort in preparation, both mental and physical, for challenging meetings is to give yourself an advantage in difficult situations. If you can practice this, you will find many situations far less threatening.

The outcome of conflict situations is often determined by the confidence shown by the parties involved. A lack of skill or knowledge can go unnoticed if you have self-assurance. A conflict can be resolved and the respect of fellow professionals earned through a display of confidence.

Self-belief and self-assurance are vital if you are to realise your potential and maximise your success at dealing with other property professionals.

Managing other people

Managing is getting paid for home runs someone else hits.
Casey Stengel (1889–1975) American Baseball Player

Managing people is something that you often think you can do, or indeed ought to be able to do. In fact some property professionals believe that you are generally employed for your skills, not your management capabilities. Should you be successful, you will be promoted. At some stage your original skills become to a large extent redundant. What you need now is management skills. But what if you haven't got those skills? This could be a very traumatic experience that threatens all your existing patterns of behaviour and lifestyle. Some people in this situation are fortunate enough to be offered management training, but the vast majority are not. You have no choice but to muddle through, copy other people or emulate a role model, all of whom may have had to do precisely that in your own past.

This chapter tackles two of the foundation stones of effective management practice: delegation and appraisals — both of which are critical to developing and improving performance in others.

Top tip for property professionals

Good management consists of showing average people how to do the work of superior people.
John D Rockefeller (1839–1937) American Industrialist

Delegation — a word that is bandied about liberally and something that people pay lip service to. People rarely delegate effectively. Mostly what happens is that you get told to do a job or take something on — that is not delegation. Very often it is an either/or situation. Either you get dumped with something you can't cope with or you don't get a chance to prove your worth because delegation is not implemented effectively. More often than not it hasn't been thought through, which results in things going wrong, breakdowns, upsets and so on. All very de-motivating for staff and project-managers alike.

As a manager you will have to assign or allocate work to others in your team(s). This will be done by balancing the work that has to be done against the availability of the other people and their abilities. Some of the work may be routine and repetitive; some may not.

Top tip for property professionals

You can delegate authority, but you can never delegate responsibility for delegating a task to someone else. If you picked the right man, fine, but if you picked the wrong man, the responsibility is yours — not his.
Richard E Krafve

When you assign work to a team member, you may retain the decision-making responsibility if it becomes necessary to adopt an alternative course of action. Delegation goes one step further and implies that the authority to make decisions is given to the team member. If you cannot delegate effectively you will find your own development will suffer and you will become snowed under with work. You need to recognise the importance of delegating work to others in your team so that you too can develop and grow.

Some reasons for not delegating

There are several reasons why some property professional managers feel reluctant to delegate. One of the most frequent excuses is: 'It's easier to do it myself.' That may be true to start with but it soon becomes a vicious circle: the more you have to do, the easier it is to do it all yourself because it is quicker than taking the time to delegate. But that road leads to overload for you and loss of morale for others. Ask yourself what might be preventing you from delegating; is it that you:

- do not understand the need to delegate
- lack the confidence with team members, and therefore, will not give them the authority for decision-making
- do not know how to delegate effectively
- have tried to delegate in the past, but failed and so will not try again
- like doing a particular job which should be delegated, but will not delegate it even though you know the team member would enjoy the job
- do not understand the management role or how to go about it
- are frightened of making yourself dispensable, so keep hold of every job
- have no time to delegate
- have nobody to delegate to.

All of these barriers need to be overcome if you are to delegate effectively.

Top tip for property professionals

Surround yourself with the best people you can find, delegate authority, and don't interfere.

Ronald Reagan, 40th US President

The skill of delegating

Delegation is a skill, like any other skill, one that can very quickly be learned. Most of it is commonsense, but here are some tips for effective delegation:

- plan delegation well in advance
- think through exactly what you want done: define a precise aim
- consider the degree of guidance and support needed by delegate
- pitch the briefing appropriately: check understanding
- establish review dates: check understanding
- establish a 'buffer' period at the end, in which failings can be put right
- delegate 'whole jobs' wherever possible, rather than bits and pieces

- inform others involved
- having delegated, stand back: do not 'hover'
- recognise work may not be done exactly as you would have done it
- do not 'nit-pick'
- delegate, not abdicate responsibility.

Top tip for property professionals

You have to do many things yourself. Things that you cannot delegate.
Nadine Gramling, American Business Executive

What should be delegated?

A manager must analyse the job he/she is actually doing in order to establish what can and cannot be delegated. You need to identify:

- totally unnecessary tasks which need not be done at all

- work which should be done by another person or in another department

- time consuming tasks not entailing much decision making which, providing training is given, could be done as well by the team member as by the manager

- repetitive tasks which over a period take up a considerable amount of time, but require more decision making and would serve to help develop a team member.

A delegation plan and timetable must then be proposed to enable time to be found to delegate. Except for the simplest of jobs, you will find that something like eight to 12 times longer will be needed to delegate a job effectively as to actually do it. However, by taking the time to delegate properly in the first place you will save yourself far more time in the future. See it as an investment in your own future as well as in the future development of the delegate.

What should not be delegated?

There are always certain tasks and authority which a manager should not delegate. This does not mean that you cannot employ staff to assist with these areas of work, but you must remain the final decision maker. These areas of work are:

- being forward looking and constantly seeking opportunities for the enterprise

- setting aims and objectives

- creating high achievement plans for your department or the part of it for which you are responsible, and ensuring quality standards are developed and maintained

- co-ordinating activity — that is knowing the task that has to be done, the abilities and needs of your own people, the resources available and then blending them to achieve optimum results

- communicating with your people and with senior managers and other property professionals and colleagues

- providing leadership and positive motivation

- the training and development of your project team

- monitoring and surveying everything that is going on and taking action necessary to maintain the planned level of achievement and quality performance.

From the above list remember that, should you delegate work which falls into any of these areas (so that you free yourself to do the jobs that you alone can do) you must remain the final arbiter. That is the action of a responsible manager.

How to delegate

Once a delegation plan has been prepared, each job must be taken separately. You must then prepare a specification which will state:

- the objective or intended goal of the job
- the method you have developed to do it
- data requirements and where the information comes from
- any aids or equipment needed to do the work
- the principal categories of decisions that have to be made
- any limitations on authority given to make these decisions; namely, when should you be consulted.

When this preliminary specification has been prepared, you must start training the delegate to do the job. Initially, close control should be maintained, but this should be loosened as soon as possible. Some form of control must be maintained, but this should not be more than is necessary to ensure that the job continues to be done properly. Keep track of which jobs you have delegated and to whom. Monitor the process with each delegate from a tactful distance.

Advantages to project team members

Delegation is often seen as being of advantage to you, the manager, but it is also of considerable benefit to the team member. The fact that jobs which you have developed are passed to others to do and the requisite authority to act is also given to them, is an aid to the development of individuals both practically and psychologically.

Delegation exercise

As you read through the list below, tick any items that you feel particularly apply to you. Then consider whether the suggested changes in what you do might be helpful.

Barriers to delegating	*How you might tackle the barrier*
• I find it difficult to ask people to do things	Try explaining to them what you will be freed to do if they take on the task.
• I do not have time to delegate	Decide to break out of the vicious circle and make time. By investing, say, half an hour explaining the task you save, say, the three hours it will take to do the task.

Barriers to delegating	*How you might tackle the barrier*
• It is quicker to do the job myself; explaining it to someone else takes too much time	It may be quicker to do the job yourself, but you have a responsibility to develop your staff members' skills. You will get quicker with practice.
• I could do the job better	Being responsible for developing the skills of your staff means investing time in development and training. In the long run it will save time. Set up a development programme.
• I need to know exactly what is happening	As a manager, you must get results through other people, or you will become overloaded — so you need to trust your staff. Build in regular feedback.
• I enjoy this job. I've always done it	As a manager, you have to let go of tasks that other staff can do. Do only what you can and should do.
• I am afraid it won't get done properly, and I will get the blame	Prepare for delegation, and build in controls as the job is done. You have the right to make mistakes.
• I am afraid someone else will do it better than I can	Set targets for your team members to do better than you at specific tasks.
• You will not do it my way	Agree to goals and targets and give freedom. There are often many ways of doing a job. A good team benefits from a variety of approaches.
• I am not sure how to do this task so feel I had better do it myself	You need to decide how to tackle the task before deciding whether it's suitable to delegate.
• The job is too big/ important	Break the job down. All jobs contain some routine elements which can be delegated.

> **Top tip for property professionals**
>
> **When in charge ponder. When in trouble delegate. When in doubt mumble.**
>
> **Source unknown**

A good delegator or a willing martyr?

Delegation is one of the most difficult things that busy property professionals need to learn. Take the test below to see what kind of delegator you are.

1. What does delegation mean to you?

    (a) Passing the buck to juniors
    (b) Dumping responsibilities
    (c) Tricking others into doing work that is rightfully yours
    (d) None of the above.

2. Are you nervous about delegating because

    (a) You do not trust anyone else to do the work?
    (b) You do not want to overburden someone else?
    (c) You have not got time to train or prepare others?
    (d) Overwork is part of your job.

3. What word would you most associate with delegation?

    (a) Risk
    (b) Fear
    (c) Guilt
    (d) Trust

4. If you did delegate a task, or tasks, would you be:

    (a) The most boring ones?
    (b) The least risky ones?
    (c) The most risky ones?
    (d) The ones that a subordinate could do just as well?

5. If you had to delegate an important job to a subordinate, would you:

    (a) Issue it as an order?
    (b) Be very apologetic?
    (c) Leave it to someone else to convey?
    (d) Present it as an opportunity?

6. When delegating to someone, do you:

    (a) Keep worrying that the job is not being done well?
    (b) Ask them to report back each time a decision is made?
    (c) Stipulate that if anything goes wrong, it is your responsibility?
    (d) Tell them only to come back to you if there is a problem they cannot handle?

Effective delegation is about trusting your staff and colleagues and delegating authority — but not responsibility. If you answered 'd' to each question, you are already a good delegator.

1. Delegation should never be forced on others, nor presented in a negative way. At best it is an opportunity for career development. However much you delegate, the buck always stops with you.

2. If you do not trust your staff, you cannot truly delegate — you will always be involved in your judgments. It is all about learning to respect and trust your team so that tasks can be more evenly spread.

3. Delegation = trust.

4. Delegating the worst jobs is not worthy of the name. You should delegate tasks that you would normally be able and prepared to do. It is not an excuse for offloading rotten jobs.

5. Good delegation needs to be presented as a positive benefit. How you 'sell' delegated tasks is most important. You should delegate the interesting and challenging jobs — and negotiate with the delegate.

6. Learn to let go. If you trust your subordinates, let them run with a task. If you feel any doubts about your capabilities, invest in training and staff development — it's cheaper than you suffering from stress-related illness.

Top tip for property professionals

When one cannot appraise out of one's own experience, the temptation to blunder is minimised, but even when one can, appraisal seems chiefly useful as appraisal of the appraiser.

Marianne Moore (1887–1972) American Poet

Performance appraisals

Appraisals are all too often the bane of a working life rather than something to look forward to and enjoy. Appraisals should be a win–win experience: both parties should gain by it and feel a sense of satisfaction and achievement. Here is an opportunity both for managers and staff to assess each others performance, build relationships and receive constructive feedback. Appraisals should be conducted once a year as an absolute minimum, with less formal quarterly appraisals in the interim.

The benefits to the individual

- Discussion of the job role in the context of job description.
- Assessing performance against agreed objectives.
- Opportunity to give and receive feedback.
- Having training needs identified.
- Opportunity to discuss career prospects and promotion.
- Future planning — understanding and agreeing objectives.
[a] Building relationships.
- Re-enforcing the delegation process.
- On-the-spot coaching.
- Increase motivation and improve morale.

Benefits to the line manager

- Evaluating performance (individual, team, organisation).
- Making the best use of resources.
- Giving constructive feedback.
- Setting and clarifying objectives.
- Identification of training needs.
- Audit of team's strengths and weaknesses.
- Receiving feedback on management style.
- Exploring and resolving problems.
- Reducing staff turnover.

Benefits to the client

- Improved performance through commitment to the project.
- A minimum standard of good project management.
- Sharing of skills.
- Appropriate manpower utilisation.
- Test of selection process.
- Reduce project staff turnover.
- Improve morale and motivation.

Preparation for the discussion

Preparation for the interview is essential if both parties are going to get the most out of it. You will need to think carefully about what you want to discuss, gather relevant information and focus on relevant issues. You will need to notify the person in writing and familiarise yourself with the individual's file and performance factors.

You will also need to think about the environment in which you conduct the appraisal. A neutral location is generally better than your office, make sure that you have what you need in the room — water, tea, coffee — ensure that you are both comfortable and that you will not be interrupted. And be sure to allow enough time.

Conducting the interview

During the interview:

- start on a positive note — emphasise what is working

- use the 10 to 1 ratio for feedback — 10 positives to 1 negative
- create a relaxed, positive atmosphere
- review the purpose of the interview
- use an agenda
- encourage the role holder to talk
- listen carefully
- use open questions
- keep to the agenda during the interview
- ensure you cover all the key aspects of the role
- discuss areas of improvement
- avoid over-criticising
- deal with one topic at a time
- summarise and maintain control throughout
- discuss further training and career development needs
- review and summarise main points, agree action plans
- end on a positive note, thank each other for your contributions.

Active listening

	Active	*Passive*
L	Look interested	Show no encouraging responses
I	Involve yourself by questioning	Ask irrelevant questions or assume
S	Stay on target	Become distracted or daydream
T	Test your understanding	Do not clarify or summarise
E	Evaluate the message	Do not connect/relate to other information
N	Neutralise your feelings	Have prejudices and make snap judgments

Constructive feedback

When giving constructive feedback remember to:

- balance praise and criticism — 10 to 1 ratio
- be constructive

- be factual and specific
- seek clarification
- maintain open communication
- focus on behaviour not personality
- be prepared to give and receive
- be honest
- agree future changes/solutions.

Appraisals can be opportunities for change on the one hand or a damaging experience on the other. Correctly used they can make the difference between a high performing and motivated individual and one that does the bare minimum to get by.

Delegation

Key action points

- Invest time in PEOPLE.
- ANTICIPATE: Strategy — competition — problems.
- THINK about: 'Tomorrow' — outstanding work — delegation.
- PLAN 'A' Time work — 4 weeks ahead.
- Establish START times.
- DELEGATE: good for you and your team.
- MEETINGS: small agenda — small attendance — precise actions — punctual finish.
- Keep your secretary/clerical support IN THE PICTURE.
- Filing system: easy RETRIEVAL.
- Keep your desk for WORK NOT STORAGE.

Appraisee's charter

I have the right to:

- receive my appraisal when it is due
- be clear what is expected from me
- have feedback on my performance
- gather my own evidence
- make genuine mistakes
- contribute equally to agreeing objectives and standards
- raise issues and concerns
- consult others.

Appraiser's charter

I have the right to:

- give feedback on performance
- contribute equally to objectives and standards
- consult others
- say 'no' to unreasonable requests
- adjourn the performance discussion
- expect certain standards of work and behaviour.

Top tip for property professionals

Without discipline there is no life at all.

Katharine Hepburn, Actress

Disciplining staff and problematic colleagues

If only your staff and colleagues worked harmoniously and efficiently all of the time? Wouldn't that be wonderful? What if they never complained, went sick or had 'attitude'? Can you imagine it? Your colleagues all resembling the characters from 'Stepford wives'? It must surely be every busy property professionals' dream. The reality however is somewhat different. People are fallible, staff and colleagues do make mistakes. Bullying, harassment and discrimination in the workplace does occur, even in the best of organisations. Unfortunately it is usually the job of the project manager to sort out the problem.

Staff and colleagues are expected to have accountability when accepting a position on a project team. They should seek to maintain the standards of work and behaviour set out under the terms of the contact, and abide by their professional code of conduct. If they trip up once or twice, perhaps a gentle reminder is all that is required. If they fail to do what is expected of them on a regular basis several adverse things happen. It costs the organisation money, upsets the balance of the project team (... if he gets away with it, why can't I? ...), morale plummets and the project manager's headaches reach epic proportions.

Everyone at work is entitled to be treated with dignity and respect. Bullying, harassment and discrimination are in no-one's interests and should not be tolerated anywhere in the workplace. Bullying is usually characterised as offensive, intimidating, malicious or insulting behaviour. Harassment is generally unwanted conduct affecting the dignity of men and women, relating to age, sex, race, disability, religion, nationality or any personal characteristic of the individual.

Top tip for property professionals

If we do not discipline ourselves, the world will do it for us.
William Feather (1888–1918) American Writer

If prevention is better than cure, one good rule for project managers is to give staff and professional colleagues examples of what is regarded as unacceptable behaviour in the circumstances. Companies whether large or small should have policies and procedures for dealing with grievance and disciplinary matters. Staff should know to whom they can turn if they have a work-related problem. Project managers should be trained in all aspects of the organisation's policies in this sensitive area.

People who are bullied or harassed may seem to overreact to something fairly trivial. However it could be the 'last straw' following a series of incidents. The dangers of allowing such behaviour to go unchecked are that they create serious problems for the organisation as a whole. These include poor morale, unharmonious staff or colleague relations, loss of respect for management, bad performance, low productivity, absences, resignations — all of which seriously damage the company's reputation.

Where a member of staff makes frequent mistakes, exhibits inappropriate behaviour, and their performance standards are falling way short of the company policy, this is not acceptable. They are showing contempt by not caring about their work, their company or the effect of their behaviour on their colleagues. Swift, decisive corrective action needs to be taken.

'Why bother?' you may ask. Why risk ending up in front of an employment tribunal, with all the concurrent hassle and traumas? That is every project manager's nightmare. Why not take the ostrich approach (... head buried in the sand ...) and hope matters improve.

Surely it's easier to leave things as they are? Well, actually, it isn't. No problem ever got smaller by leaving it alone. What starts as a minor dispute can develop into a full-blown crisis if ignored. By not confronting the issue, things only get worse.

Top tip for property professionals

Error is the discipline through which we advance.
William Ellery Channing (1780–1842) American Writer

Discipline should not be confused with punishment. Discipline is positive; punishing someone is to do with exacting a penalty. Effective discipline involves dealing with the shortcoming or misconduct before the problem escalates. Disciplining a member of staff or colleague can be an informal or formal procedure, depending on the severity of the problem. If an informal approach is appropriate, counselling or training can provide a vital role in resolving complaints. Whichever the case, the important thing is to follow a fair procedure. When the issue involves a complaint about bullying, harassment or discrimination, there must be fairness to both the complainant and the person accused.

Set a good example: the faster you deal with the problem the stronger example of management behaviour you are giving. If you allow time to elapse between the incident taking place and disciplining the member of staff or colleague, the message you are sending is that you couldn't be bothered to do much about it. Maintain fair procedures for dealing promptly with complaints from staff and colleagues. Set standards of behaviour by means of an organisational statement to all staff or through the company handbook. Finally, let staff and colleagues know that complaints of bullying, harassment or discrimination will be dealt with fairly, confidentially and sensitively.

Because of recent changes in the law, the prudent manager should consult the *ACAS Advisory Handbook: Discipline and Grievances at Work*. The *ACAS Code of Practice: Disciplinary and Grievance Procedures* gives advice on good practice in disciplinary matters. This is something which is taken into account in cases appearing before employment tribunals.

First of all, identify the problem. Is the issue you are dealing with related to performance or misconduct? Try to deal with the matter as quickly as possible. Make sure you explain why the individual is being disciplined. Describe exactly what the unacceptable behaviour is. Here it is essential to avoid generalities, be as specific as possible. Stick to the facts; focus on the behaviour and not the person. Explain the effect their behaviour or actions are having on the rest of the company/department/unit. Specify what changes need to be made and outline the consequences if the unacceptable behaviour continues.

In order to comply with the law, which requires fairness above all things, you will need to carry out a full investigation, giving the individual the opportunity to state their case. They are allowed to be accompanied to any interview or hearing by a colleague or company representative. Make sure you give an explanation for the disciplinary action and specify clearly the appeals procedure.

In the case where you suspect someone has made an unfounded allegation of bullying, harassment or discrimination for malicious reasons, it is essential such allegations be investigated fully and dealt with fairly and objectively under the disciplinary procedure.

Depending on the outcome of the disciplinary procedure, reasonable action should be taken in relation to the facts. Penalties are not always necessary, sometimes it is more appropriate to offer counselling or training. In the severest cases, where bullying, harassment or discrimination amounts to gross misconduct, dismissal without notice may be the right course of action.

Where such issues arise, project managers should examine their company policies, procedures and working methods to see if they need to be improved. Useful contacts where further advice can be sought include:

- Advisory Conciliation and Arbitration Service   *www.acas.org.uk*
- Commission for Racial Equality   *www.cre.gov.uk*
- Disability Rights Commission   *www.drc.org.uk*
- Equal Opportunities Commission   *www.eoc.org.uk*

In conclusion, should you, as a project manager, have to discipline a member of your staff or report a professional colleague, remember the golden rule: keep records of everything. You cannot have too much documentation when taking personnel actions. Give fair warnings (along with notification of the consequences) and always in plenty of time. It is also important to give your staff or colleague enough time to respond, or rectify their behaviour. Make sure your company's policies are reasonable and that standards are achievable. Finally, make clear what avenues exist for appeal and that the staff or colleague knows what they are.

The law in this area is complicated and all employers contemplating dismissal, or action short of dismissal, such as loss of seniority or pay, are required to follow a three-step statutory procedure. For more details see *www.dti.gov.uk* or consult your employment lawyer for specific advice.

Recruiting and selecting the right people

He who would search for pearls must dive below.
John Dryden (1631–1700) British Poet and Dramatist

Finding and hiring the best applicant for a job is no easy task. With lots of people looking for work, it is challenging to have to pick the best person from a large number of candidates. Whether you are about to hire your first employee, or you have taken on staff many times before, you know the feeling — it is a leap in the dark. Recruitment and selection is a vital task which managers frequently have to fulfill. Get it right and everyone benefits. Get it wrong and the consequences are dire.

Any ambitious property professional who wants their department or project team to grow, sooner or later needs to take on staff. There comes a time when outsourcing has reached its limits, there is no time to finish the countless tasks, more hands are needed without doubt. To make the task seem less daunting (taking into account the growing complexity of the government's employment laws) you need to proceed with care.

Take a look at the various steps involved in the recruitment process.

Defining the job to be done

Analysing the job and drafting the job description. The creation of a clear and tangible job description is an essential first step. Investing time at this stage is a good policy. Whether the position is a new one or you are filling an existing one, before starting the recruiting process, be sure you know what standards you are going to use to measure your candidates.

Write down the description of the job, whether it is a newly created post, or an existing position. What is the job title? What are the objectives and purpose of the job? What duties, responsibilities and tasks go with it? How does it fit with existing jobs? Where will it lead and what prospects can new staff be offered? Describe the reporting lines and working relationships. State the specific tasks, standards and responsibilities required. Detail the appraisal procedure and be clear as to the remuneration package and other benefits.

While a clear job description is fundamental to successful recruitment, the personal profile sets out the characteristics of the kind of person who might be qualified and suited to undertake the role. People are the core of any business and never more so than in the property and construction industry. Some people love their jobs and others live for their work.

Top tip for property professionals

Actors search for rejection. If they don't get it they reject themselves.
Chevy Chase (1943–) American Actor

Specifying the profile of the likely candidate

Identifying the characteristics of the person who is most likely to be suitable for the position is useful. Descriptions such as hard working, good attitude, experienced, stable, smart and responsible spring to mind. But how do you find such paragons?

Personal characteristics

This covers age, qualifications, experience and special skills. Basic personal characteristics such as age, education, experience, specialist qualifications — for example fluency in a foreign language. The purpose here is to make the selection process manageable. Most employers wish to trawl a fairly wide area, but they are not keen to plough through hundreds of applications, some of which are unsuitable.

Character traits

Do you want someone creative, industrious, loyal or innovative? Aspects of character, such as these, are important attributes but much more difficult to measure accurately.

Motivational factors

Will the job suit someone who wants a steady routine or someone who wants something more challenging? The manager needs to look here at what is likely to appeal to an applicant about the job. Is it suitable for someone who is ambitious, competitive, innovative and creative? These are set as a guide only.

Responsibility

Areas of responsibility relate to the aspects of character which make a candidate suitable for the post. Does the applicant have the ability to work on his own, care for others or give presentations to large audiences? Will they need to work as part of a team? Is 100% accuracy essential in their work?

The worst case scenario is to end up appointing someone who proves not able to do the job but not so bad that he can be sacked. It is important to consider the kind of person you feel best suited to the position. Once you have given some thought to these details, you can start the selection process.

Top tip for property professionals

A man travels the world over in search of what he needs and returns home to find it.

George Moore (1852–1933) Irish Writer

Sources of candidates and methods of attracting the right person

From among the sources for attractive potential candidates, perhaps the most effective is internal selection, but there are other options.

Internal selection

Via the HR or Personnel department, internal advertisement of the position. Provided an effective training and selection programme is in place, it is possible to source and select for the new position from within the company. The advantage here is that the applicant is known to the manager, the applicant knows the company and has 'bought-in' to the culture of the firm. It is good for staff morale to see that inside promotion is possible.

Referrals

If you are looking to fill a vacancy, make sure you let people know. Whether it is co-workers, colleagues, friends, relative or clients — many good candidates are sourced from referrals. Someone whom you know can give you great insights into the applicant's strengths and weaknesses and character. You will get far more information than you would from resumes alone.

External advertising

If you are writing a job advertisement, make sure the copy describes the actual job to be done, and describes the organisation in terms of what it does and its style and culture. It also needs to state clearly a specific salary range and the nature and qualifications of the candidates sought. Situations vacant advertisements are relatively inexpensive and can get your job publicised over a wide area. This may have its advantages but the disadvantage is that you may have to sort through literally hundreds of applications to find a few good ones to shortlist.

Temporary agencies

If in doubt, hire a temp or locum. This will give you some relief if the work is piling up while the recruitment process is underway. It also provides the opportunity to try out employees before you hire them. If you like your temp, ask the agency if you can hire them for a nominal fee or after a certain period of time.

Recruitment consultants

These can be used for sourcing applicants for a specialised position or if you simply do not have the time to go through the whole process of recruitment and selection yourself. The agency carries out the advertising, recruiting and screening of applicants, providing you with a shortlist of perhaps five people to interview.

Executive selection and headhunters

The higher the level of the position you are seeking to fill, then it may be appropriate to seek assistance from one of the executive search companies of headhunters. They have great experience and are able to select candidates to the highest standard.

The Internet

It is possible to use the Internet as an effective advertisement. Web pages allow you to present large amounts of information on your company and your job opportunities. The internet is available 24/7 and reaches a huge audience.

Assessing written personal details

When considering applicants' CVs bear in mind they can be quite distinctive.

A targeted CV draws attention to the applicant's skills and focuses on the qualities that make him the right person for the particular position.

A chronological CV is one that summarises the qualifications and career experience of the candidate. This form of CV is popular with local authorities, central government and more traditional employers.

An experience-based CV is valid for individuals who work in specialised areas of employment. It describes their track record to date.

CVs should be precise and list a candidate's achievements with detailed points. Beware CVs which generalise — this could indicate a weak candidate.

There is no need for large amounts of personal detail to be added to a CV. It should be a piece of personal marketing literature — focusing on the product — in other words — the skills the candidate offers.

CVs which are gimmicky are risky. A CV should look attractive, clean and professional. Trendy typefaces, coloured ink are not appropriate. If a CV contains spelling mistakes, it could indicate a careless individual.

Be vigilant at checking accuracy of CVs — employment history, qualifications and skills.

Top tip for property professionals

Never hire anyone who is going to report directly to you who you do not intuitively just plain like from first impressions. If your instincts tell you you're going to have a hard time working with someone, pass.

Fred Charette

Systematic approach to interview

When it comes to interviewing candidates, you do need to ask the right questions. Because hiring the right people is essential to the growth and success of your business, the manager needs to get his interview techniques right. This means asking loaded questions which will reveal the information you need to make an informed decision.

At the outset, welcome the applicant, then begin by summarising the position. Ask the prepared questions and use the candidates answers to evaluate his strengths and weaknesses. Conclude the interview after allowing the candidate the opportunity to ask any questions he or she wishes. Advise them when you will be making your selection.

Take time to prepare your questions and make notes of the applicant's answers.

'Why are you here?'

This may elicit the answer 'because I want a job with your firm'. But it could give you other information, which you would never have gleaned, had you not asked in the first place.

'What can you do for us?'

An important question and one which candidates should be prepared for. It is quite common for applicants today to approach the recruitment process on the basis of 'what can your company do for me?' and this question redresses the balance.

'What kind of person are you?'

You need to know — after all, if recruited, you will be spending considerable periods of time in their company. You want to employ someone who will be congenial most of the time he/she is at work.

'If you stayed with your current company, what would be your next move?'

This is a question designed to extract information on several levels. It should reveal a sense of what the applicant expects but also why the applicant wants to move on. If, for example, the applicant says he wants to be a manager but the person above him has been there for 25 years, you can move on with the interview. If he were to say that he hoped to be promoted within six months, why is he leaving that job? You should try to elicit the real reason why he wants to leave that company.

'What do you consider makes you exceptional compared to others?'

A difficult question for most people, because applicants have a tendency to be uncomfortable praising themselves. If the answer is given in a reasoned manner, the applicant may have a good degree of self-esteem and some courage. A timid response could indicate a reticent type — not one fit for a challenging role within your company. Beware the applicant who launches into a lengthy monologue about why the world revolves around him/her. The over-active ego could spell disaster if this applicant is aiming to fill a position where teamwork is a requisite.

'Describe your greatest achievement to date'

If the applicant can recall quickly and with detail a satisfying and recently accomplished project, which he/she recounts in a measured, comprehensive way, you may have a winner here. Any applicant who is quick enough to think on his/her feet and produce the anecdote without hesitation is likely to be an asset to your company.

'Do you need many hours a week to get your work done?'

This question is designed to elicit the work ethic of the potential employee. If he/she expects to put in long hours with your company, this could indicate that he/she is staying late to do extra work, or merely that he/she works inefficiently. A discussion as to working habits can reveal how he/she will fit in with the rest of the employees. A company where it is normal to stay until 7pm would not suit someone with a strict 8 to 5 mentality.

'What sort of salary are you expecting?'

There is no point in having gone through the selection process only to reach the end of the interview and find out that your idea of a competitive salary and benefits package is so far removed from the candidates that you seem to be on different planets. You may not be able to offer enough in terms of salary, but try putting together a generous benefits package — including pleasant office, corporate membership of a health club, impressive job title, generous holidays, health insurance, pension etc.

Don't forget to keep copious notes as you interview each of the candidates. It will be impossible to remember who said what and your written notes will be an essential aid when evaluating the applicants.

There are one or two topics on which it is not appropriate to question applicants: these relate to the applicant's race, skin colour or national origin. Also anything to do with marital status, religion, and criminal record. Personal details such as height, weight, financial status and disabilities are also not necessary. These questions do not relate to how the applicants perform their jobs and are best avoided.

Assessment and checks

Check, check and check again. References, skills, previous employment history. Do the dates on the CV match with other facts? Was that extended holiday actually spent at Her Majesty's pleasure?

It is quite surprising how many people exaggerate their education experience. This is a good place to start your checks. If it is found that the application is inaccurate about one thing, it is likely that the rest of their CV is too.

Don't be hesitant about calling previous employers for information about an applicant. You should get more detail from a supervisor or manager than from the HR or Personnel department, who is more likely simply to confirm the dates the applicant was employed.

With regard to skills, if the job requires good presentation skills, ask the applicant to make one. If they need to write well, look at some examples. If the requirement is for fluent French, set a test for the applicant.

Top tips for property professionals

The work of science is to substitute factors for appearances, and demonstrations for impressions.

John Ruskin (1819–1900) Critic

Final selection and appointment

When reviewing the information about all the candidates, you will be glad to look at the notes you made. (You did take notes, surely?) How does each applicant stand up against your original criteria for the position? Is there an outright winner? How many losers? Sort them into categories: Winners, possible winners and losers. Be objective. Don't be influenced by irrelevant elements such as clothes or hairstyles.

If there is more than one candidate with equally good qualifications, it may be necessary to go for a second or third round of interviews. This may seem time consuming or expensive, but it is better than making a wrong appointment and living to regret it.

If you really can't be sure — go with your gut instincts. Although they may seem equally matched in skills and experience, you will probably have a feeling that one is more suitable than the other. If so, allow your intuition free reign.

It is sometimes wise to hire people for their personalities rather than for their skills and qualifications. Whereas it is difficult to change someone's personality, it is not impossible to teach them new skills or train them in certain techniques.

If in doubt, don't be afraid to ask advice. Managers hiring staff must be certain of the decision they are about to make. Ask your mentor, or an experienced director of a well-established company. Departmental managers cannot afford to make mistakes and most people are happy to help if requested to do so.

Never hire someone on the basis that 'he's the best of a bad bunch'. This is potentially disastrous. Better to repeat the whole process again because rarely do people make a miraculous change for the better once they are appointed.

Once you have made your decision, telephone the successful applicant as soon as possible and offer your first choice the job and secure his acceptance. If he is no longer available go to candidate B. Hopefully you will be able to hire someone from among your selection of 'winners'.

Don't forget to communicate with the unsuccessful candidates. It is old-fashioned courtesy, but it costs nothing to be polite and it will make them feel a lot better. A short letter to them saying you will consider alternatives (if they might suit another position in the company) or keeping their CV on file (and telling them you are doing so).

Finally, don't forget that if your department or team grows the way it should, your new employee could be a manager one day. They will be running the office in your absence. By looking for someone who possesses that extra spark, someone who would be happy to take responsibility and act on their own initiative, you could be choosing your successor!

Motivating
the team

Motivation is the art of getting people to do what you want them to do because they want to do it.
Dwight Eisenhower (1890–1969) 34th US President

Motivation matters. There is considerable research to show that people, who are, to put it simply, happy in their work, will perform better than those who are not. Many busy property professionals have, as part of their job, the need to get results through other people, rather than for them. If this is so in your case, the motivational state is important. In terms of both productivity and quality of action maximising motivational feeling will assist performance.

Similarly, it is easy for any dilution of motivation to act to reduce performance; something that ultimately reflects on a team manager. Multiply the effects, either positive or negative, by the number of people reporting to you and you see the real importance.

You as project manager must act not just to ensure that people perform well on their particular project, but that they do so consistently and reliably. Good motivation also acts to make sure that people are as self-sufficient as possible, able to make decisions — good decisions — on their own and take action to keep things running smoothly. If you have to check every tiny detail and issue moment by moment instructions, then neither productivity nor the quality achieved are likely to be as good as they might be. There is every difference in the world between people being able to do something and do it well, however, and being willing to do it and do it well.

Property professionals who are managers need to motivate people, rather than leave them to their own devices. Motivation, like so much else in management, does not just happen. It must be recognised as an active process — one that you need to allow some time for on a continuing basis.

Top tip for property professionals

The most important thing about motivation is goal setting. You should always have a goal.

Francie Larrieu Smith, Runner

The fundamentals of motivation

The theory of motivation is extensive and this is not the place to do otherwise than recap some essential principles (if you are entirely familiar with the principles, then by all means skip to the next main heading).

The essentials of motivating successfully

Many people, certainly in years gone by, took the view that getting performance from staff was a straightforward process. You told them what to do, and they did it. If that was, for some reason, insufficient then it was backed by the power of management; effectively by coercion.

Management by fear still exists. In any economy with less than full employment the ultimate threat is being out of a job. But whether the threat is subtle or specific, whether it is just an exaggerated form of arm-twisting or out and out bullying, even if it works (at least short term) — it is resented. Your job as a manager is not simply to get things done, it is to get things done willingly. The resentment factor is considerable. People fight against anything they consider to be an unreasonable demand. So much so that the fighting may tie up a fair amount of time and effort, with performance ending up as only the minimum people 'think they can get away with'.

If people want to do things and are encouraged to do things well, only then can they be relied on to actually do them really well. Motivation provides reasons for people to want to deliver good performance.

If this sounds no more than common sense, then that is because it is. For example, are you more likely to read on if you're told that if you do not, someone will come round to your house and break all your windows? Or if you're persuaded that you will find doing so really useful and offer you some sort of tangible reward? (It is intended that you will find it useful, incidentally, but sadly there is no free holiday on offer.) Motivation works because it reflects something about human nature, and understanding the various theories about this is a useful prerequisite to deploying motivational techniques and influencing staff behaviour.

Top tip for property professionals

Motivation is simple. You eliminate those who are not motivated.
                    Lou Holtz (1937–) American Football Coach

---

Theory X and Theory Y

The first of the classic motivational theories that is worthy of some note was documented by Douglas McGregor. He defined the human behaviour relevant to organisational life as follows:

- Theory X: makes the assumption that people are lazy, uninterested in work or responsibility and thus must be pushed and cajoled to get anything done in a disciplined way, with reward assisting the process to some degree.

- Theory Y: takes the opposite view. It assumes people want to work. They enjoy achievement, gain satisfaction from responsibility and are naturally inclined to seek ways of making work a positive experience.

There is truth in both pictures. What McGregor was doing was describing extreme positions. Of course, there are jobs that are inherently boring and mundane, and others that are obviously more interesting and it is no surprise that it is easier to motivate those doing the later. Though having said that, it is really a matter of perspective.

There is an old, and apocryphal, story of a despondent group of convicts breaking rocks being asked about their feelings concerning the backbreaking work. All expressed negative feelings, except one — who said simply 'it makes it bearable if I keep the end result in mind — I'm helping to build a cathedral'.

Whether you favour Theory X or Y, and Theory Y is surely more attractive, it is suggested that motivation creates a process that draws the best from any situation. Some motivation can help move people from a Theory X situation to a Theory Y one. Thereafter it is easier to build on positive Theory Y principles to achieve still better motivational feeling and still better performance; and your communication should reflect this fact.

Top tip for property professionals

The mechanics of industry is easy. The real engine is people, their motivation and direction.

Ken Gilbert

Maslow's hierarchy of needs

Another theory that helps describe the basic situation on which all motivational effort must be directed is that of Abraham Maslow. He wrote that peoples' needs were satisfied progressively. In other words, only when basic needs are met do their aspirations rise and other goals set.

The first such needs were psychological: enough to eat and drink, warmth, shelter and rest. In a working environment people need to earn sufficient money to buy the answers to these factors. Next come needs of safety and protection: ranging from job security (one that is less easily met than once was the case) to good health (with the provision of health care schemes by employers now very common).

Beyond that he described social needs: all those associated with working in groups and with people. The work environment is a social environment, indeed for some people contacts formed through work may represent the majority of the people contact in their lives. Linked to these are a further level of needs such as recognition within the organisation and among the people comprising the work environment, and the ability to feel self confidence, self fulfilment and look positively to a better future, one in which we are closer to realising our perceived potential and happier because of it.

However you define and describe this theory, it is the hierarchical nature of it that is most important. What it says is that peoples' motivations can only be satisfied if this hierarchy is respected. For instance, it suggests that motivational input is doomed to be ineffective if it is directed at one level when a lower one is unsatisfied. It is thus little use to tell people how satisfying a job is, if they are consumed with the thought that the low rate of pay makes them unable to afford basic essentials. Thus all communication with staff designed to have a motivational impact must bear in mind the whole picture.

Top tip for property professionals

People say that motivation doesn't last. Well neither does bathing — that's why we recommend it daily.

Zig Ziglar

Hertzberg's motivator/hygiene factors

This last theory leads to a view of the process that links much more directly to an action based approach to creating positive motivation. Hertzberg described two categories of factor: first, the hygiene factors — those dissatisfiers that switch people off if they cause difficulty. And second, the motivators, factors that can make people feel good.

Dissatisfiers or hygiene factors

- The dissatisfiers (or hygiene factors): these he listed, in order of their impact, as follows:

    - company policy and administrative processes
    - supervision
    - working conditions
    - salary
    - relationship with peers
    - personal life (and the impact of work on it)
    - status
    - security.

All are external factors that affect the individual (because of this they are sometimes referred to as *environmental* factors). When things are without problem in these areas, all is well motivationally, if there are problems they all contain considerable potential for diluting any positive motivational feeling.

Note: it should be noted here, in case perhaps it surprises you, that salary comes in this list. It is a potential dissatisfier. Would you fail to raise your hand in answer to the question: would you like to earn more money? Most people would certainly say 'yes'. At a particular moment an existing salary may be acceptable (or unacceptable), but it is unlikely to turn you on and be a noticeable part of your motivation. So too for those who work for you.

It is, for instance, things in these areas that give rise to gripes and to a feeling of dissatisfaction that rumbles on. If the firm's parking scheme fails to work and you always find someone else in your place, perhaps someone more senior who it is difficult to dislodge, it rankles and the feeling is always with you.

There are many things springing from these areas for managers to work at. Getting them right can make a positive contribution to boosting the motivational climate.

The restriction here is that these things are not those that can add powerfully to positive motivational feeling. Get things right here and demotivation is avoided. To add more you have to turn to Hertzberg's second list.

Top tip for property professionals

Motivation is like food for the brain. You cannot get enough in one sitting. It needs continual and regular top ups.

Peter Davies

Satisfiers or motivators

These define the key factors that create positive motivation. They are, in order of relative power:

- achievement
- recognition
- the work itself

- responsibility
- advancement
- growth.

It is all these factors, whether positive or negative and stemming from the intrinsic qualities of human nature, that offer the best chance of being used by management to play their part in ensuring that people want to perform and perform well. Communication is a vital part of this picture. Every piece of communication can have motivational overtones — and probably will. For example, put in a new system, say asking people to fill in a new form on a regular basis, and, if it is not made clear why it is useful, people will be demotivated (because it relates to the list of dissatisfiers — specifically policy and administration — above).

Similarly, a wealth of different communications all affect the motivational climate, moving the overall measure one way or the other. For example:

- *job descriptions, clear guidelines and adequate training*: all give a feeling of security, without which motivation suffers

- *incentives*: will work less effectively if their details are not clearly communicated (for instance, an incentive payment scheme may seem so complicated that no one works out how they are doing and motivation suffers as a result)

- *routine jobs*: can be made more palatable by communicating to people what an important contribution they make

- *job titles*: may sensibly be chosen with an eye on how they affect peoples' feelings of status as well as acting as a description of function ('Sales Executive' may be fine and clear to customers, but most prefer titles like 'Account Service Manager').

Furthermore, the same essential act can be changed radically in terms of the effect it has motivationally just by varying the way in which communication occurs. For example, the simplest and least expensive positive motivational act you, as manager, can engage in is probably uttering the simple phrase 'Well done' (and which of you can put a hand on your heart and say you do even that sufficiently often?)

Consider some different ways of doing it, listed in what is probably an ascending order of motivational power:

- Saying well done, one to one
- Saying it in public, maybe in an open plan office
- Saving it at an 'occasion' (anything from a departmental meeting to a group taking a coffee break together)
- Saying it (in one of the ways listed) and then confirming it in writing
- Getting the initial statement (however it may be done) endorsed by someone senior
- Publishing it (eg in a company newsletter).

The implications here are clear. Not only is motivation itself primarily executed through communication, but the precise form of that communication needs to be born in mind and contributes directly to the effect achieved.

Top tip for property professionals

We know nothing about motivation. All we can do is write books about it.
Peter F Drucker (1909–) Management Consultant

Producing positive results

It may seem from what has been said already that motivation is a complex business. To some extent this is so. Certainly it is a process affected by many, and disparate, factors. The list of factors affecting motivation, for good or ill, may be long, and that is where any complexity lies, but the process of linking to them in terms of action is often straightforward.

The very nature of people, and how their motivation can be influenced suggests five important principles for the manager dedicated to actively motivating people.

1. There is no magic formula

No one thing, least of all money, provides an easy option to creating positive motivation at a stroke, and anything that suggests itself as such a panacea should be viewed with suspicion.

2. Success is in the details

Good motivation comes from minimising the factors that tend to create dissatisfaction, and maximising the effect of those factors that can create positive motivation. All of them in both cases must be considered; it is a process of leaving no stone unturned. Anything contributing to the overall picture should be utilised.

What is described as the motivational climate of an organisation, department or office is the sum of all the pluses and minuses of each individual factor.

3. Continuity

The analogy of climate is a good one. A greenhouse is a small-scale example of this. Many factors contribute to the temperature inside. The heating, windows, window blinds, whether a door or window is open, if heating is switched on and so on. But some such things — whatever they are — are in place and contributing to the prevailing temperature *all the time*.

So too with motivation. Managers must accept that creating and maintaining a good motivational climate takes some time and is a continuous task. Anything, perhaps everything, they do can have motivational side effects. For example, as was mentioned, a change of policy may involve a new system and its use may have desirable effects (saving money say). But if complying with the system is seen as bureaucratic and time consuming the motivational effect may be negative, despite the change having positive results.

Overall the trick is to spend the minimum amount of time in such a way that it secures the maximum positive effect.

4. Time-scale

Another thing that must be recognised is the differing time scales involved here. On the one hand, signs of low motivation can be a good early warning of performance in peril. If you keep your ear to the ground you may be able to prevent negative shifts in performance or productivity by letting signs of demotivation alert you to the coming problem.

The level of motivation falls first, performance follows.

Similarly, watch the signs after you have taken action aimed at actively affecting motivation positively. Performance may take a moment to start to change for the better, but you may well be able to identify that this is likely through the signs of motivation improving. Overreacting because things do not change instantly may do more harm than good.

If motivation is improving, performance improvement is usually not far behind.

So, the timing of communication is vital too. A busy moment and something allowed to go by default may lead to problems at some point in the future.

5. Bear others in mind

There is a major danger in taking a censorious view of any motivational factor — positive or negative. Most managers find that some at least of the things that worry their staff, or switch them on, are not things that would affect themselves. It is the other people who matter. If you regularly find things that you are inclined to dismiss as not of any significance, be careful. What matters to you is *not* the same as what matters to others.

If you discover something that can act for you influencing your people, however weird or trivial it may seem, use it. Dismissing it out of hand — and in communications terms, say, failing to explain something adequately — just because it is not something that you feel is important will simply remove one factor that might help influence the motivational climate. It will make achieving what you want just a little more difficult. At worst, it will also result in your being seen as uncaring.

Similarly, what is important to you may not be to others. This is an important factor that any manager forgets at their peril. A further aspect of motivation now needs to be added, that concerned specifically with involving people.

Involving people

The word empowerment enjoyed a brief vogue in the mid-90s, as one of a succession of management fads that, if you believe the hype, solve all problems and guarantee to put any organisation on the road to success. If only that were the case. On the other hand, there is sense in the idea of involvement which is essentially the meaning of empowerment. It may not solve everything, but it is useful and it does provide additional bite to the prevailing motivational feeling.

> **Top tip for property professionals**
>
> **The difference between involvement and commitment is like ham and eggs. The chicken is involved; the pig is committed.**
> *Martini Navratilova (1956–) US Tennis Player*

Empowerment in action

Rather than describe leaden definitions, let us start with an example. The Ritz-Carlton have enjoyed good publicity not only for the undoubted quality of their many hotels, but for a particular policy they operate. Say you are staying in one of their hotels and have (perish the thought) something to complain about. So, reeling from the stench from your mini-bar or whatever, you stick your head out of your room door into the corridor and take up the matter with a passing chambermaid.

Now whomever you were to speak to the procedure would be the same. Every single member of the hotel's staff is briefed to be able to handle your complaint. They do not have to find a supervisor, check with the manager or thumb through the rulebook. They sort it. As they think fit. And they have a budget to do so — every single member of staff can spend so many dollars (it started as something like $500, but has no doubt changed) instantly, and without any checks, to satisfy a guest's complaint.

So, to continue our example, if the mini-bar was dirty they could summon someone to clean it at once (even if that meant paying overtime) refill it with complimentary drinks and throw in a free bottle of fine wine and a bowl of fruit on a side table to make up to the guest for the inconvenience.

Such staff are certainly empowered.

It is an approach that gets things done. It regards staff as a key resource, not only one to get tasks completed but one who can, in many ways, decide just how they get it done. The empowerment approach goes way beyond simple delegation and plays on the appeal of responsibility to the individual to get things done and done right. It works in part because staff like it: being empowered is motivational.

Behind empowerment

On the other hand, empowerment does not allow managers to abrogate their responsibility, nor does it represent anarchy, a free for all where anything goes. The chambermaid (mentioned above) does not have the right to do just anything, only to select, or invent, something that will meet the customer's needs and which does not cost more than the budget to implement.

Staying with our hotel example, consider what must lie in the background, staff must:

- *understand guests*, their expectations and their likely reaction to difficulties (and how that might be compounded by circumstances — having to check out quickly to catch a flight, for example)

- *be proficient at handling complaints* so as to deal with anything that might occur promptly, politely and efficiently

- *have in mind typical solutions* and be able to improvise to produce better or more appropriate solutions to match the customer situation

- *know the system*: what cost limit exists, what documentation needs completing afterwards, who needs to be communicated with etc.

The systems — rules — aspect is, however, minimal. There is no need for forms to be filled in in advance, no hierarchy of supervisors to be checked with, most of what must happen is left to the discretion of the individual members of staff.

The essence of such empowerment is a combination of self-sufficiency based on a solid foundation of training and management practices that ensure that staff will be able to do the right thing.

Letting go

Often, on training courses, the room is full of managers tied, as if by umbilical cords to their mobile telephones or pagers. Many of the calls that are made in the breaks are not responses to messages, they are just to 'see everything is all right'. Are such calls, or the vast majority of them, really necessary?

The opposite of this situation is more instructive. See if this rings a bell. You get back to the office after a gap (a business trip, holiday, whatever). Everything seems to be in order. When you examine some of the things that have been done you find that your view is that staff have made exactly the right decisions, yet ... you know that if you had been in the office, they would have asked you about some of the issues involved. Some of the time staff empower themselves, and when they do, what they do is very often right.

All empowerment does is put this kind of process on a formal footing. It creates more self-reliant staff, able to consider what to do, make appropriate decisions and execute the necessary action successfully.

Perhaps you should all allow this to happen more often and more easily.

Top tips for property professionals

Power can be taken but not given. The process of the taking is empowerment in itself.
                              Gloria Steinem (1934–) Feminist Writer

---

Making empowerment possible

Empowerment cannot be seen as an isolated process. It is difficult to view it other than as an integral part of the overall management process.

You can only set out to create a feeling of empowerment by utilising a range of other specific management processes to that end. The process perhaps starts with attitude and communication. What degree of autonomy do your staff feel you allow them? If they feel restricted and, at worst, under control every moment of the day they will tend to perform less well. Allowing such feeling is certainly a good way to stifle initiative and creativity.

So you need to let it be known that you expect a high degree of self-sufficiency, and manage in a way that makes it possible. All sorts of things contribute, but the following — all aspects of communication — are certainly key.

Clear policy

Empowerment will only ever work if everyone understands the intentions of the organisation (or department), their role (clear job descriptions) so as to allow them to put any action they may need to decide upon in context.

The other requirement of an empowered group is an absence of detailed rules to be followed slavishly, but clear guidelines about the results to be aimed at.

Clarity of communication

This has been mentioned before, but is especially important in the context of motivation. Any organisation can easily be stifled by lack of, or lack of clarity in, communication; an empowered group is doubly affected by this failing.

Little interference

Management must set things up so that people can be self sufficient, and then keep largely clear. Developing the habit of taking the initiative is quickly stifled if staff know nothing they do will be able to be completed without endless checks (mostly, they will feel, made just at the wrong moment).

Consultation

A management style in which consultation is inherent acts as the best foundation for an empowered way of operating. It means that the framework with in which people take responsibility is not simply wished, perhaps seemingly unthinkingly, upon them, but is something they helped define — and of which they have taken ownership.

Feedback

Empowerment needs to maintain itself, actions taken must not sink into a rut and cease to be appropriate because time has passed and no one has considered the implications of change. Feedback may only be a manifestation of consultation, but some controls are also necessary. Certainly the overall ethos must be one of dynamism, continuing to search for better and better ways to do things as a response to external changes in a dynamic, and competitive, world.

Development

It is axiomatic that if people are to be empowered, they must be competent to execute the tasks required of them and do so well. Remember too that useful development is itself always a significant motivator.

An enlightened attitude to development is motivational. A well-trained team of people are better able to be empowered, they have the confidence and the skills. An empowered and competent team is more likely to produce better productivity and performance. It is a virtuous circle.

Top tips for property professionals

Most teams aren't teams at all but merely collections of individual relationships with the boss. Each individual vying with the others for power, prestige and position.
Douglas Murray Mcgregor (1906–1964) Management Theorist

Achieving the right balance

At the end of the day the answer is in your hands. Keep too tight a reign on people and they will no doubt perform, but they may lack the enthusiasm to excel. Management should have nothing less than excellence of performance as its aim — market pressures mean any other view risks the organisation being vulnerable to events and competitive action.

On the other hand, too little control, an abrogation of responsibility and control, also creates risk. In this case that staff will fly off at a tangent, losing sight of their objectives and, at worst, doing no more than what takes their fancy.

Like so much else a balance is necessary. Empowerment is not a panacea, but an element of this philosophy can enhance the performance of most teams. Achievement and responsibility rank high as positive motivators, and empowerment embodies both. Motivation will always remain a matter of detail, with management seeking to obtain the most powerful cumulative impact from the sum total of their actions, while keeping the time and cost of so doing within sensible bounds.

Empowerment is one more arrow in the armoury of potential techniques available to you, but it is an important one. Incorporate it in what becomes the right mix of ideas and methods for you, your organisation and people, make it clear to people how you operate and it can help make the whole team work effectively.

Finally, here are ten ways of achieving a motivational style for success:

1. always think about the people aspects of everything
2. keep a list of possible motivational actions, large and small, in mind
3. monitor the 'motivational temperature' regularly
4. see the process as continuous and cumulative
5. ring the changes in terms of method to maintain interest
6   do not be censorious about what motivates others, either positively or negatively
7. beware of panaceas and easy options
8. make sufficient time for it
9. evaluate what works best within your group
10. remember that, in part at least, there should be a 'fun' aspect to work.

Make motivating, and the communication that transmits it, a habit. Take a creative approach to it and you may be surprised by what you can achieve with it. The motivation for you to motivate others is in the results.

Aiming for excellence

Finally in this chapter, remember that even the best performance can often be improved. Motivation is not simply about ensuring that what should happen happens. It is about striving — and achieving — excellence. All sorts of things contribute, from the original calibre of the staff you recruit to the training you give them. Motivation may be the final spur that creates exceptional performance where there would otherwise only be satisfactory performance.

It is an effect worth seeking; and it is one multiplied by the number of staff involved. How much more can be achieved by ten, 20 or more people all trying just that bit harder, than one manager, however well intentioned, doing a bit more themselves? Motivation makes a real difference.

Mentoring

Mentoring can help development and act quickly to improve performance, and thus enhance career progress.
Patrick Forsyth, of Touchstone Training & Consultancy

The workplace is not an entirely benign environment; indeed the phrase 'corporate jungle' has been used of it with some truth. Of two things you can be sure. First, organisations are dynamic. Things change and so too must people: they need old skills enhancing, new skills adding and new ways of looking at and dealing with things. Development is not an attractive luxury — it is an ongoing necessity. To succeed in any organisation you need to change and adapt along with the circumstances.

Second, development does not just happen. No individual employee can assume that development will be automatically provided by their employer, certainly not all that is desirable. Everyone benefits from taking an active approach to development. Employers should think constantly about what might help employees and staff need to see some of the responsibility for prompting action as their own.

A gallimaufry of methods should be used — and such range widely from simply reading a book (or an article) to attending a course or logging onto some kind of computer-based training. This chapter deals with one, sometimes underestimated, method: that of mentoring. This is the technique of creating a relationship, often separate from normal reporting lines, where one person helps another and does so on a continuing basis, with much of what happens conducted on an informal basis.

Recruit a mentor

One of the motivations for being a manager, or at least for some managers, is the satisfaction of helping people develop and of seeing them do well. A number of successful property professionals have been lucky in that early on in their careers people they worked for took on the role of mentor to them. Today, they owe much to these people. It is something well worth aspiring to, whether to mentor others (if you are in the position of giving something back) or seeking a mentor from among those with whom you come into contact. At best you learn a great deal and very much quicker than would otherwise be the case. On the other hand, luck does play a part here. If there are not suitable candidates around, not everyone has the courage or opportunity to seek out such assistance, particularly early on in their careers.

The ideal mentor is sufficiently senior to have knowledge, experience and clout in your chosen profession, property, construction and engineering. They need to believe in the process, for it to appeal to them, and they need to have time to put into the process. Such time need not be great. The key thing is to have the willingness to spend some time regularly helping someone else. If your own boss and your mentor is one and the same person, that might be ideal, but it is not essential. Indeed some would define a mentor as someone who is specifically *not* your boss. Usually, if such a relationship lasts, it will start out one way — they help you — but may well become more two-way over the years; perhaps the person on the mentor side makes the decision to help rather on the basis of this anticipated possibility.

After the first few years of your property and construction career, there is no reason why you cannot have regular contact with a number of people where in each case the relationship is of this nature. This can take various forms depending on the area of expertise of the individual. For example if you wish to benefit from a number of different areas of expertise, you may have to seek out a number of different individuals. Such an arrangement is not uncommon.

Mentoring explained

Perhaps a little explanation as to exactly how mentoring works. It can be so useful. A mentor is someone who exercises a low key and informal developmental role. More than one person can be involved in the mentoring of a single individual. While what they do is akin to some of the things a line manager should do, as has been said, more typically in terms of how the word is used, a mentor is specifically *not* your line manager. It might be someone more senior, someone on the same level or from elsewhere in the organisation. An effective mentor can be a powerful force in your development. So how do you get yourself a mentor?

In some organisations, and it is common practice in the property, construction and engineering profession, this is a regular part of ongoing development. You may be allocated one or able to request one. Equally you may need to act to create a good mentoring relationship. You can suggest it to your manager, or directly to someone you think might be willing (or persuaded) to undertake the role, and take the initiative.

What makes a good mentor? The person must have authority (this might mean they were senior, or just that they were capable and confident). They should have suitable knowledge and experience, counselling skills, appropriate clout and a willingness to spend some time with you (their doing this with others may be a positive sign). Finding that time may be a challenge. One way to minimise that problem is to organise mentoring on a swap basis: someone agrees to help you and you line up your own manager to help them, or one of their people.

Then a series of informal meetings can result, together creating a thread of activity through the operational activity. These meetings need an agenda (or at least an informal one), but more important they need to be constructive. If they are, then one thing will naturally lead to another and a variety of occasions can be utilised to maintain the dialogue. A meeting — followed by a brief encounter as people pass on the stairs — a project and a promise to spend a moment on feedback — an email or two passing in different directions — all may contribute. The topic of these encounters may be general — a series of things all geared, let's say, to improving presentations skills; these might range from help with preparing to a critique of a rehearsal. Or they may focus very specifically: aiming to ensure that a particular report being written is well received.

Thus the meetings that constitute mentoring activity can:

- be planned and long term
- link to other activity (everything from a course attendance to job appraisal)

- be opportunistic, scheduled at short notice to tie in with events
- consist of one exchange or a number of linked ones.

An eye must be kept on the overall continuity, with no long gaps for instance and with the time scheduled conveniently for both parties so that it does not become a burden.

Ultimately what makes this process useful is the commitment and quality of the mentor. Where such relationships can be set up, and where they work well, they add a powerful dimension to the ongoing cycle of development, one that it is difficult to imagine being bettered in any other way.

This kind of thing should be regarded as really very different from, and very much more than, networking. The nature and depth of the interaction and the time and regularity of it is much more extensive. This is not primarily a career assistance process in the sense of someone who will give you a leg up the organisation through recommendation or lobbying, though this can of course occur. It is more important in helping develop the range and depth of your competences, improving your job performance with this in turn acting to boost your career. A senior or well-connected mentor may also act as an early warning system too when trouble is brewing.

As a final note, the perfect mentor is someone who makes an art form of this kind of role. Mentoring is a development technique that can assuredly make a difference, and a very positive one at that. Job performance and career progress can both benefit directly.

Changes and going forward

In a time of drastic change it is the learners who inherit the future. The learned usually find themselves equipped to live in a world that no longer exists.
Eric Hoffer (1902–1983) American Author and Philosopher

Project Group Ambiguity

In any organisation confusion reigns where there is board level ambiguity. If colleagues compete for their views to prevail, take for example a partnership in a professional service practice, then those below them will be confused.

If they converge and present a united front, then those below will be motivated by clear messages rather than demotivated by those that are imprecise or ambiguous.

This general rule applies equally well to property professionals working together in a project team. Should one of them fail to recognise the importance of presenting a united front, the delivery of a successful project becomes much more difficult.

Leadership

On any project team, whether it comes from the titular leader(s) or not, there has to be a 'champion' who lives and embodies the new 'dream'. If everyone is supposed to be responsible, but no one in particular is responsible, there will be a dissipation of focus, effort and achievement.

Single goal

A clear and sustained purpose should be identified at the outset to which people can commit. It needs to be understandable and upliftingly relevant at all levels in the group.

Clarity of purpose

There has to be a clear reason for the project, and should any changes in style or attitude have to take place, this has to be accepted by the group or else inertia can set in or worse, dictatorial or unhelpful behaviour.

The illusion of unity

Don't expect everyone to back the project changes, but don't drop to the lowest common denominator of agreement either. You need the stimulation of sceptics who will either be right or force a repeated re-evaluation of the adopted changes.

How open should you be?

Tell people as much as practicable. Take some risks with candour — this is likely to work better in a group whose change delivers progress than one facing adverse circumstances.

Communication

Effective communication is vital and almost impossible to overdo. (This topic is referred to many times in this book.) Yet communication links can break down in stressful situations due to being given inadequate priority. Uncertainty and major change create anxieties that must be assuaged if morale and performance are not to slump.

The rule of proportionate responsibility

The more senior you are, the more responsibility you must take. Every top management action sends a signal. Delegation cannot deflate into abdication if an important change is to progress effectively.

Teams and leaders

Good teams and good leaders support each other. Teams often contain hierarchies of power and responsibility. Senior people may not lead the team but they have a powerful sponsorship role in the organisation.

Structure and culture

Use structure to change culture. It is normally easier and faster to change the structure, reward and measurement systems and the performance criteria than to try to resculpt the mists of an elusive team or group culture.

Creating winners

Personal success is a great motivator but a punitive regime fosters risk aversion and a blame culture. It is better to recognise meritorious failure as well as success to show that those in the project team are valued.

Fast change and initial acts

Early successes create productive momentum. There should also be intermediate milestones of achievement that raise people's confidence — not only in the project — but also in their ability to succeed further. Speedy successes (or quick hits) even in relatively minor matters, demonstrate purpose and commitment.

Caring for casualties

Caring for people, as well as being morally commendable, is organisationally effective. The worse you treat those who are no longer valued enough, the more resistance to change will grow. Survivors will adopt defensive strategems in case they may be next.

Minimising unintended consequences

You cannot avoid all errors, but you can organise yourself to anticipate some and recover from others. Mainly this will be to do with contingency resources which a well-laid plan can allow for.

Top tip for property professionals

If we don't change, we don't grow. If we don't grow, we aren't really living.

Anatole France (1844–1924) French Writer

Going forward

Much more could be said about the benefits of holistic working practices within the property and construction industry. When considering the industry as a whole, it matters little whether you are involved in large project teams and departments, or small specialist groups, working harmoniously with others is always most desirable.

If you can achieve this most of the time you will be scaling the ladders and avoiding the snake pits while you progress your career.

What has been covered in this book shows that when working with other people, respect their differences and be courteous and polite wherever possible. Bad manners does not help to keep a diverse group of individuals united, however important the project may be.

Top tip for property professionals

Human relationships always help us to carry on because they always presuppose further developments, a future ... and also because we live as if our only task was precisely to have relationships with other people.

Albert Camus (1913–1960) French Existential Writer

If you can develop the skill of problem solving, negotiation and persuasive tactics, dealing with other people — whether they be property professionals or lay clients — you will achieve more with less stress and effort. One of the advantages of working in the property and construction industry is that you are going to be dealing with lots of different types of people. Some of them will have huge egos and others will be diametrically opposite and lack confidence. Whichever level of the assertiveness scale these people occupy — and some can be off the scale — it is essential to recognise who and what you are up against and temper your own behaviour accordingly. The combined skills of an acrobat, a diplomat and a doormat are frequently called for. Should you possess these — all you have to do is work out which one is needed, when and to what degree. Simple, isn't it?

Should you be in the position of having to manage others, whether it is in a department or in a project group, listen to them and inspire them. Motivated individuals are happy and will work better and be more productive. Project teams need to be able to deliver: for that to happen a united and positive attitude among team members is essential.

There is no doubt the importance of clear communication is paramount. Where there is ambiguity or doubt progress cannot be made. Much has been written about communication styles and methods and it is worth paying attention to this should you feel your skills are weak in this area. It is only by persuasiveness can you move projects forward should the road ahead be strewn with rocks and boulders. Smooth operators go far and will take others forward with them. Look out for any you meet and emulate them when opportunities occur.

Top tip for property professionals

Technological progress is like an axe in the hands of a pathological criminal.
Albert Einstein (1879–1955) German Physicist

In the chapters dealing with technology, it is sensible to make sure that whatever technology you use is fit for purpose. There are many time and labour saving devices available to everyone — pay attention to what you are doing. Just because swift duplication and dissemination of information is available at the press of a button, don't over-use a system simply because it is there. 'Need to know' is fine but make sure you know who does and who doesn't.

People matter — there's no doubt about it. People unlike tasks cannot always be hurried. Short cuts are fine but when dealing with project colleagues, staff and clients if time is needed to keep everyone happy, make sure you devote that time. Crucial among property and construction professionals is the need to be respected and supported. The projects are so important they cannot be left at risk because someone can't be bothered to ask, listen or understand someone else's point of view.

Top tip for property professionals

Those people who develop the ability to continuously acquire new and better forms of knowledge that they can apply to their work and to their lives will be the movers and shakers in our society for the indefinite future.

Brian Tracy, Trainer, Author, Businessman

If you are ever in any doubt about how to deal with someone, just think about how you would wish to be treated should the roles be reversed. Whether it be recruiting, selecting, mentoring, motivating, appraising, promoting, getting it right is essential. Get it wrong and it can take ages to recover from very costly mistakes. Team building is a science in itself and good project managers know this only too well.

Behaviour and performance are important to any professional whatever industry they work in. In property and construction it is easy to become highly stressed and over worked because of the pressures on teams to deliver sophisticated and time sensitive projects. Take good care of yourself, as well as looking after those with whom you work. Stress up to a point is positive, but over a certain level it becomes dangerous. Time management skills and personal effectiveness among individuals are highly desirable qualities to any client or employer. Keep your skills in good repair and your career will progress well. Take advantage of any training courses that are offered along the way — there is always something that can be learned or reviewed.

Part 5

Appendix
Case Studies

**Contributed by
Tanya Ross**

Managing project managers

Project Manager — one of those terms that means different things to different people. The Association of Project Managers defines project management as:

> The person with authority to manage a project. This includes leading the planning and the development of all project deliverables. The project manager is responsible for managing the budget and workplan and all Project Management Procedures (scope management, issues management, risk management, etc.)

This is a pretty comprehensive definition. It sounds as though the project manager is responsible for everything! A client who has a large portfolio of projects, or perhaps a lay client may choose to delegate the day-to-day business of running projects to someone else, perhaps someone with previous experience and particular skills in this area. In this way, inexperienced or over-busy clients remove themselves from the confusion of the myriad decisions that need to be made every day in order to maintain progress. The role of client representative is created, investing responsibility in someone who may be more familiar with the construction process, entrusting them with keeping an eye on things and dealing with the nitty-gritty decisions. Often this role may be taken by a quantity surveyor (on the basis that costs are close to the client's heart) or perhaps a 'professional' project manager. The project manager is there to protect the client's interests and to ensure that the project is delivered according to plan: on time, on budget and to an appropriate quality.

How the project manager achieves their goals will depend upon the particular contractual arrangements, the nature of the project, and, to a large extent, on the personalities of the individuals involved. In practice, there are three broad categories of project manager:

- the client's rottweiler
- the paperwork king
- the design integrator.

The client's rottweiler

An ambitious developer has a number of key contracts ongoing, and has drafted in a manager with a reputation as an achiever to deliver a substantial city centre project. The manager has over 20 years experience in the industry, and prides himself on getting things done to the fastest possible programme. Happily, the client lets him get on with it, content that he has removed himself from the tedious meetings that the architect insists on having, and the endless lists of 'Decisions Outstanding' or 'Information Required'. The project manager rubs his hands in glee, delighted to get the chance to boss around this signature architect. He boasts down the pub that he's got 'Sir Pompous' under control, and that all it takes is someone with some balls in order to get these effete architectural types to toe the line.

Initially the project moves forward fast, with the early packages coming forward ahead of schedule, and the project manager feels justified that his bombastic style is bearing fruit. However, as things move on, it becomes clear that the relationship between the project manager and the rest of the team is far from harmonious and this begins to affect the progress of work. Since the project manager does not have a deep understanding of the various elements and how they interface — he relies on a simple linear programme which demands release of design information in a pre-determined sequence. This does not allow for design development between packages. Problems emerge between different packages where the details haven't been sufficiently thought-through. Worried that he might not achieve his deadline (and the hefty bonus attached) the project manager becomes increasingly strident in his criticism of the designers. They protest vigorously that of course things are not fully co-ordinated because they did not have enough time in the unrealistic programme to sort out all the details. And by the way, the façade cladding will need to be re-ordered because the signature architect insists on a particular shade of terracotta and has persuaded the planners that this is the only solution. So there!

At this point, things can go two ways: relationships deteriorate to such an extent that someone walks away from the project; or the developer gets involved and starts throwing the threat of lawsuits around in an attempt to achieve a reconciliation. Either outcome is less than ideal for the project: better to foster constructive relationships between the manager and the team members that he is aiming to direct, and put the project ahead of any personal agendas.

The paperwork king

At the outset of a large complex project, a project manager is appointed by the client to deal with the day-to-day running of the job. The manager, with a background as a quantity surveyor, is determined to get a system in place that permits the monitoring of all decisions, and ensures that the client signs off any changes. (He's been stung before: accused of spending his client's money unnecessarily and he's not about to undergo that excruciatingly uncomfortable dressing-down again!). He introduces a 'change control system', which consists of a series of forms that need to be completed when any deviation from the baseline design is proposed. In practice, this sounds like a sensible idea, and should allow everyone to keep a record of any alterations requested and ensure that the whole team is aware of the current proposals. There's a 'proposed client change' form (PCC), initiated by the client — or the architect on his behalf. Then there's a 'proposed design change' form (PDC) initiated by the design team. Finally there is a 'proposed buildability change' form (PBC) initiated by the contractors team. Each change has to be evaluated in cost terms, time terms, and health and safety terms.

Gradually, it becomes clear that a small team is needed within the contractor's team and the project manager's team in order to keep up with the evaluation process. There is a considerable amount of robust debate about what constitutes a change: every re-located door and partition that improves the spatial arrangement? Surely that is just 'design development'? The project manager spends a chunk of his time in 'refereeing' disagreements between the contractor and the design team about whether every drawing amendment constitutes a material change. The quantity surveyors in each team spend considerable time arguing with their opposite number about the value of proposed changes, such that a whole day every week gets consumed with confrontation as individuals tussle over the value of four number M16 bolts over six number M12 bolts ... Although the project appears to be under control, and the client is delighted that he's appointed such a conscientious manager, inexplicably, progress is not as swift as anticipated, and the contractor seems to be asking for more money for the most trivial of decisions. The client begins to get uncomfortable and issues an ultimatum: no more changes!

Unless the team manages to shake off the weight of paperwork under which it has been buried, there is a danger that progress grinds to a halt: the contractor refuses to accept any more drawing alterations, and the architect refuses to give approval to construction information based on 'unacceptable' designs. A Mexican standoff ensues, where each party blames the other for cessation of progress, and the only apparent solution is for the client to throw money at the project.

The design integrator

A rare breed, the design integrator is a project manager that understands that design is not a linear process, but rather a cyclical one, with each cycle converging towards the optimum solution. Such a manager will permit a certain design latitude, providing time in the programme for design finalisation, for co-ordination with other packages (and other designers) and will try and align tender dates for collections of packages together. This approach, while it may find favour with the design team, may cause friction among the contracting team, as it marches relentlessly towards a series of imposed deadlines. The trick is to find an acceptable compromise between design time (and the benefits that this can bring to the ultimate product) and tender deadlines.

In practice, tender deadlines are set on a somewhat spurious programme string, which gives a generous margin for mobilisation and lead-in. A little interrogation will usually establish more realistic target dates: does the painting package really need a 32 week mobilisation period? With non-fabrication packages, there should be no reason to require anything other than a few weeks for sorting out the order. Having said that, there will always be market fluctuations. In a particularly heated market, it is possible that fit-out contractors (for example) may be quoting a 12 week mobilisation period; for items manufactured in Europe beware the summer close-down which may dictate factory dates; for specialist items, such as specialised joinery or glass, expect to manage a long lead-in period. With good communication between project manager, design team and contractor's procurement department it should be possible to dovetail reasonable design deadlines with practical procurement periods, without incurring additional costs.

Dealing with
lay clients

'Right, what I want is a building,' the client states confidently. 'Here, on this site.'

'Umm ... and?' queries the architect.

'Er ... please?' suggests the client.

I suspect that what the architect was really after, was more information rather than a polite addition. It's great to have a confident client, ready to build, but a considerable amount of decision making has to be done before the design team can really start work on formulating a building design.

For a start, is a new building really the solution to the client's problem? A simple lack of space can be overcome in more imaginative ways: an addition to an existing building; an internal re-planning exercise; decentralisation of the business itself perhaps. It's all too easy to spot an empty neighbouring field and seize on the possibility of an ego-satisfying shiny addition to the building stock. Perhaps the more sustainable solution is to establish how existing resources can be harnessed rather than indulging in the expenditure of new resources. That's not to say that construction advice is inappropriate — far from it — but experienced professionals should be able to explore alternatives, advising clients of real costs and genuine business options. Once the client is sure that a new building is the optimum solution, then they can confidently move onto the next stage.

In these case studies he and she are used interchangeably and in each case may be assumed as referring to both he/she.

'OK, now I know I want a new building. I can't fit all my people in the existing one and there's no economic alternative to something new. Here. Please.'

Not so fast: there's still some more work to be done before we start designing. The next stage is establishing a brief. How big does he want the building? What purpose will it serve? How will it be used? What are the budget limitations? When does he want it finished? Without clear unequivocal answers to all these questions, the design team don't have a hope in delivering an appropriate solution.

Take the first question: how big? The client may be able to tell you 4,500m², but then again he may not. He may be only be able to say how many people he wants to contain. Then he will need assistance in translating the number of people into a net area and a gross area. And of course the ratio between net and gross can be a significant factor in assessing a building cost and footprint. £1,200/m² seems like a reasonable rate for construction, but applied to a net area or a gross area? And what net to gross ratio is reasonable anyway? The difference — even with modest building footprints — can be a sizeable suitcase of notes. Getting it clear right at the start will save huge arguments and acrimony later.

Then there is how the building is to be used. The client may not have considered this yet, but the design team will need to think about whether it's used at night, if the occupants are sedentary computer users, how people arrive. The more these things are defined, the better the design will suit the users. Of course, if the development is speculative, and its eventual use is unknown, then the design team have a different problem: the main driver is maximum flexibility. In some ways this is easier to deal with: flexibility rules every design decision from floor to ceiling height to the colour of the walls. It's imperative that the whole team understand the motivation for building, what is really important to the client: is it money, is it noise, is it time, is it eco-friendliness, is it corporate aesthetics?

There is a terrible temptation to give an inexperienced client the building you think he ought to have, rather than the building he actually wants. Exploiting his inexperience may result in a beautiful sculptured edifice, much-photographed and applauded, but it may not necessarily result in a functional space where the kitchen is actually next door to the dining-room. It's my belief that most clients would prefer to have a hot supper than a design award.

Contractors too are guilty of exploiting woolly briefs. If the client doesn't quite know what he wants, any reasonably switched-on contractor is going to suggest the bath with a bigger mark-up, a cheaper brick at no extra cost, a less labour-intensive joinery package ... So while this may give the contractor some much-needed profit margin, it might not necessarily give the best product for the job. A tighter brief could help limit such abuses.

So it's up to designers to be responsible in educating clients, to help them develop a consolidated clear brief, with as little ambiguity as possible, that is really going to get them the building they need. This may take time, but it's time well-spent, that will save everyone in the long run. From a well-written brief, it's much easier to develop a design, much quicker to arrive at solutions, as many of the clients key decisions have already been taken. Here's to an end to woolly briefs and a welcome to the contented client.

While there are a number of repeat clients out there — the government in all its many guises; the supermarket chains; major developers — a significant number of construction projects every year are carried out for clients who have little or no knowledge of the construction industry. These 'lay' clients have a different approach to commissioning building projects from experienced repeat clients, and the canny property professional will recognise the need to treat them differently too.

Chances are, that if you have a lay client, it's because they're building a 'one-off'. Perhaps it's a new wing for the museum; a new stadium; a substantial public amenity: whatever the reason, it's the only building that they're likely to commission in their time. Each of these clients wants their new facility to be the best it can be and expect to get the professional team to deliver just that. They don't really want to know about the intricacies of the building process, but they do want to see their vision realised in bricks and mortar (or glass and steel, as it may be).

The museum client

Years of dedicated fund-raising have allowed the museum to commit to construction of a new wing. A long-cherished project, the museum director is terrifically excited that a major donor has come forward just at the right time, so that she has a chance to influence how the museum can house its precious collection of medieval artefacts. Allied to this, the donor has an interest in education and is keen for learning facilities to be included in the new project to promote an understanding of history for the local school children. Between them, the museum director and the benefactor will guide the development of the project and make the necessary decisions.

As the project is launched, a cost consultant is appointed and architects and engineers are invited to make proposals. At this point, the only guidance that the teams have is a budget — a suspiciously exact budget at that. It transpires that the client has taken the fund-raising target, knocked off 17.5% for VAT and 13% for fees and arrived at a number. Without knowing any of this, the design teams do their creative best, and proffer exciting sketches that whet the appetite, encouraging the client to believe that they can afford five new galleries rather than three; that a sculptured glazed entrance is really going to put them on the map and that the innovative sun-tracking system is the latest in cutting-edge sustainable technology. The clients are delighted, and really enter into the spirit, suggesting that an iconic building should be a key feature of the brief, and that sustainability should underpin the whole design. As the teams are appointed, and the project unfolds, it becomes clear that the brief is considerably at odds with the 'real' construction budget. Hey presto! Before the team has really started, the client is being told that they can't have what they thought they wanted.

Such a scenario is not unusual. If lay clients are not given (or do not seek) sound cost advice at the outset of a project, then disappointment is inevitable. And some of the arcane oddities of construction will be a complete mystery to them: what is a 2.5% Main Contractor's Discount? Why do I need to keep something back for the Defects Liability Period? And why do I need a contingency? This last question is, perhaps, the most crucial of all. As experienced professionals, we know that things can go wrong. Sometimes it's the weather (too windy for the tower crane to operate); sometimes it's the unexpected on site (unrecorded Saxon burial ground anyone?) and sometimes it's just plain human error (the fork-lift driver didn't mean to reverse into that glazed panel). No construction project should start without a decent measure of contingency: and the more complex the project, the more substantial the contingent sums need to be.

As the project continues, the client gets over their initial disappointment, and begins to engage with the appointed designers to shape the new building. Quickly, they decide where they want their own offices to be, they decide upon a tiered auditorium for the education centre, they decide upon a particular marble, from a particular quarry, for the gallery floors. But there are dozens of other decisions to be taken every day, and the museum director doesn't have the requisite knowledge to make all of these. Why should she know the acoustic reverberation time for the auditorium? Or the loading requirement in the delivery bay? Faced with this lack of knowledge, the design team will need to make some decisions on behalf of their client. This is fine — a competent architect, supported by capable engineers and specialists should have no difficulty in suggesting the most appropriate answer for all these day-to-day questions. However, it is a wise team that records all such decisions clearly so that if there is a query somewhere down the line, the reasoning behind such decisions, and indeed the acceptance of such decisions on its behalf by the client, is clear and unambiguous.

The sports club client

A series of successes in national competitions has raised the profile of a local sports club, and the resulting surge in membership has prompted them to launch into an ambitious expansion and re-location plan. A past president has some construction experience, well, he's a retired estate agent, anyway — so he is entrusted with the redevelopment plan. At the outset, things go well — their existing city centre site attracts the interest of house builders and the club is able to generate a substantial sum from the sale of the existing pitches. Committed to moving on, the club engages a high-profile architect together with a reputable team of designers and cost consultants to deliver their vision of a new state of the art stadium, complete with spa and sports medicine facility and space for the media that will surely flock to the new club.

Initially the project proceeds well, and work soon starts on the new site. However, as the new ground takes shape, it emerges that the simple, yet substantial club-house envisaged has evolved into a clever piece of architectural detailing, and co-ordination on site is proving tricky and time-consuming. Soon, the programme is two months behind, and nerves begin to set in: the house-builders are eager to start digging up the old pitches, and the new clubhouse has to be ready in time for the beginning of the next season. Faced with this looming deadline, the client representative decides that he needs to spend more time getting involved, and starts to micro-manage the project. He insists on being involved in every decision, summoning the design team to interminable meetings to scrutinise everything that's been done in the week and instituting a paper trail for even the tiniest of changes.

While keeping records is undoubtedly a key requirement of any construction project, and quality assurance systems are instilled to ensure that records are kept, there is a point where excessive paperwork can really slow a project down. One half of the office-based staff on any building job are there to ensure the smooth logistics of the site — ordering materials, organising deliveries, ensuring the right staff turn up to work — while the other half are there simply to keep track of the changes. Changes are the things that make money for contractors, but they're also the things that cost a job time. A simple system that records changes and the reasons for them is invaluable in sorting out final accounts, but such a system can become over-complicated. Too many pieces of paper, or approvals required before authorisation, and suddenly a tiny change ('We'll have the cherry veneer rather than the beech veneer for the reception desk please') becomes a hugely time-consuming operation, when a simple phone call to the supplier might do the trick just as well. Of course it's a balance, and the most successful projects tend to be those where systems are clear and simple and where the client trusts his designers and builders to make at least some of the decisions on his behalf.

The public amenity client

An impoverished local authority wins a bid for lottery-funding for a new 'attraction' to be constructed in a derelict industrial park, with the aim of injecting some new life into a rundown and deprived part of the county. Delighted to have some cash to allocate, the authority asks one of their long-standing councillors to head up a build committee. The councillor takes her responsibilities seriously, anxious that every decision is reached with consensus from members.

Slowly, the form of the attraction emerges from committee meetings: it will be educational, have an element of thoughtful spirituality — recognising the multi-faith community — and it will be fun! Such requirements in one amenity prove a challenge for the designers, but gamely they struggle on through three different concepts designs and then four different scheme designs. Time doesn't seem to be an issue with this project, which is just as well. Instead, consultation the watch-word: community workshops, questionnaires and public meetings ensure that everyone has a chance to voice their opinions. When the project finally reaches the point of actually constructing something, four years have passed, and the serious-minded councillor unexpectedly loses her seat in the local elections. Bemused, the client body appoints a new chair of the build committee who decides that the whole project needs to be reviewed before any more money is spent, given that half the allocated award has already been spent on fees and consultants before a single brick has been bought. The design team, already exhausted by the constant re-design, finally reach the end of their tether and resign en masse, much to the bemusement of the client body.

Whenever a project is run by a 'committee', experienced professionals sigh and add half a percent to their fees. No matter how well-meaning, decisions by committee take five times as long and are always couched in so much vague language, that the designers are forced to interpret what might be required and inevitably discover that it's 'not quite what we were after.'

Getting recognised

A consultant came to visit the other day. This is not an unusual occurrence. Consultants often come to visit, to sell their services in one way, shape or form. Methods of selling have become increasingly sophisticated — less the dubious raincoated door-to-door salesman, more the suave suited service provider that you can't do without. Anyway, this particular consultant had come to advise us on corporate development. Now, I'm not entirely sure I know what corporate development is, or whether we need it, let alone how much it should cost.

The consultant was terribly plausible, had read all those management tomes (you know the ones — orange covers and far too many words in the title: 'The Seven Habits of Good Managers' or 'Fifty Ways to Turn Your Business Around — or Your Money Back!') and nodded sympathetically as I explained our approach to learning and development. He had good credentials, an engineering degree and ten years in the construction business, but — and this is the bit that really got to me — his business card described him as 'Bob Smith, MBA' (OK, the name's been changed to protect the innocent). Now, I'm sure he put in considerable effort (and expenditure) to acquire his MBA, but why did he choose to ignore his engineering qualification? Four years of unmitigated hard work, and somehow he's embarrassed about admitting to an MEng? It's a very good qualification. It demonstrates numeracy, analytical ability, logical thinking and a certain dogged perseverance in order to complete 30 hours-a-week when contemporaries in humanities are cruising around campus with eight hours of lectures per week.

It's symptomatic of engineering's lack of self-worth — no other discipline seems to be ashamed at admitting to post-nominal letters. The Communications Director at the Engineering Council has confessed that 'engineers do have a tendency to be rather self-effacing', readily ducking into the shadows if the spotlight of attention happens to turn in their direction. And there are a lot of engineers within the construction industry: civil engineers, structural engineers, electrical engineers, ground engineers, fire engineers, etc., all of them apparently suffering the same reluctance to admit to any pride in their profession and in what they do.

And it seems to be endemic across the industry — construction has such a poor image that we only sheepishly divulge our occupation. I've heard a quantity surveyor describe himself as 'an accountant for buildings' — good grief! We're trying to impress by calling ourselves accountants?! OK, I'm prepared to reveal that I have been known to describe myself at parties as 'in publishing' to avoid the glazed and slightly panicked look that invariably comes over a stranger's face when confronted by the truth.

So, enough of that! We should be prepared to stand up and be counted, to state boldly 'I'm an engineer in the construction business.' It's something to take pride in: look around at the extraordinary achievements in construction in the last five years and register how we've improved our world. Whatever their architectural merit, the combination of lottery funding, imaginative designers and committed constructors has resulted in world-beating buildings like the Millennium Dome, the Eden Project, the Millennium Bridges in Gateshead and over the Thames (yes, yes, but they sorted out the wobble), and the Glasgow Wing Tower to name but a few.

By refusing to acknowledge our contribution, we are perpetuating the myth that construction is a dirty, messy business that's a necessary evil within the economy. The image is difficult to refute when the popular media enjoy stories of 'Builders from Hell!', and 'Why Buildings Fall Down.' Influencing the media approach is a challenge to all of us, and there's a terrible temptation to wring our collective hands bemoaning that 'no-one really understands us'; but the first step is to recognise that we're doing good stuff. Then we need to have coherent and sympathetic representation to project into the broader public arena. One or two industry figures are already doing this (Mark Whitby springs to mind), but we need more — many more. Preferably young, articulate, enthusiastic and confident in talking about what they do. And yes, we need to be prepared to admit what we do at parties.

It is a challenge we should relish, for personal as well as professional reasons. Whether it is not having to explain to your granny exactly what it is you do for a living, or being considered suitable matrimonial material, even a modicum of increased public awareness could mean improved recognition. We are real people in this industry, doing real and valuable work, contributing to both the financial well-being and visual environment of the whole country. If we want to be recognised as such, we must be prepared to be more visible, to step out of the shadows and tell the public what we do, unashamed, unabashed and uninhibited.

Getting things done

Passing the Buck ...

Picture the scenario (I'm sure you've been there a thousand times):

Contractor's surveyor : 'Ere, they've forgotten to give a slab level on this construction drawing, but it scales as 12.75, I'll set out at that shall I?

Contractor: Excuse me, Mr Engineer, can you tell me the structural slab level?

Engineer: Er, sorry ... I'm waiting for the architect to confirm the finishes so we know what allowances to make from the finished floor level. Mr Architect?

Architect: Oh dear, I can't tell you yet, I'm waiting for the client to confirm what level of use he is anticipating, and for the interior designer to confirm that it's available in his preferred colourway. Just let me check with the client.

Client: Our corporate colour is green, but our Facilities Manager is very keen not to have carpet and wants to use this new lino he saw at a conference in Holland. Oh, and I can't tell you the long-term use of the space as we're only leasing the building for two years — you'll have to check with the landowner ...

One simple question, and 17 different people need to be involved in the decision.

This is perhaps, an overly simplistic example, and yet it illustrates the hierarchy of decision-making that can lead to the lengthy periods of time between the issue of an apparently straightforward 'Request For Information' and the receipt of an answer.

So here's the alternative scenario:

Contractor's surveyor : 'Ere, they've forgotten to give a slab level on this construction drawing, but it scales as 12.75, I'll set out at that shall I?

Contractor: Yes, but I'd better check. Excuse me, Mr Engineer, can you confirm the structural slab level is 12.75mAOD?

Engineer: Yes (thinks : that gives the architect 50mm to play with for his finishes, that'll have to do — he'll probably never notice anyway, and I can always claim the contractor didn't comply with the concrete tolerances if the edges don't work out OK).

Still not terribly realistic perhaps, and I would hope that the engineer does at least manage to have a constructive conversation with the architect rather than make the snap decision and just hope for the best.

Sometimes the ripples from the stone of that snap decision can lap on the shores of far distant ponds. There was the junior architect, entrusted with handrail details, who decided that a particular dimension of stainless steel rail would look fetching. At stairs, the rail had to turn down the stair, but because there was a minimum bending radius to the rail, the stairs had to be re-set out slightly differently. Because the stairs moved, the associated lifts had to move so that the platform interface would still work. Because the lift moved, the lift pit had to move. The lift pit was already constructed and cost the client £15,000 to move.

So how are decisions extracted? Who decides who has to make the decision? Invariably it will depend on the individuals involved — some are only happy when issuing instructions left, right and centre, some will only make a decision after prolonged cogitation and consultation. Again, it's a question of balance. It can become a question of delegation too.

If I had a pound for every time someone had told me 'Effective managers delegate', I'd be able to be retired and living in Notting Hill. But there has to be a point at which delegation stops. I remember vividly one occasion where an electrical engineer was asked to write a specification, but since he was pressed for time, he asked a junior colleague to dig out the last specification that was written, and amend it to suit the current project. The young engineer, intimidated by the size of the document, didn't actually read it, but asked a secretary to find the document and change all references to the project title. The spec was produced in time, but it wasn't until tenderers started asking questions that it was realised that the spec appeared to be for a computer centre rather than for the small nursing home intended.

So delegate yes, but maintain control. Ask advice yes, but within reason. 'A problem shared is a problem halved.' 'Two brains are better than one.' There are any number of adages in support of sharing problems, and any industry which involves creative design and evolving technologies can only operate effectively when knowledge is pooled.

But sharing a problem doesn't necessarily mean that responsibility for solving that problem is passed on. A recent survey by British Telecom on the uses and abuses of e-mail, identified that business users are increasingly using e-mail as a form of anonymous psychotherapy. Sharing difficulties with colleagues, unburdening themselves of apparently insoluble problems with user-groups, not to mention sharing a taste for appalling jokes and risque screen-savers with friends across the world. Telling someone else about a crisis may make us feel a lot better, but that doesn't bring a solution any closer. Increasingly, construction professionals are choosing to pass the buck and ask someone else the difficult question rather than bite the bullet and come up with an answer.

There has to be an acknowledgement of responsibility when decisions are made, and there has to be sufficient information for those decisions to be made responsibly.

Delivering on promises

How many times have you heard the excuse 'Oh, yeah, sorry, it's late — it'll be with you on Monday instead.'? It's become almost as much of a clichè as 'The cheque's in the post'. There's an increasing level of non-performance, coupled with indifference, as the people doing the job seem unwilling or unable to deliver on promises made.

The design team agrees to the project manager's programme, with an issue date on a Friday. Mentally, the architect is allowing himself the weekend to finish off the drawings, reasoning that missing the post on Friday is an entirely plausible and unpunishable offence. The drawings don't quite get finished as planned and, by the time all the copies have been printed, the issue sheets have been completed correctly and the post boy has found enough A3 envelopes, the vital drawings don't reach their intended destination until the following Thursday — nearly a week later than that pyramid-shaped blob on the project manager's sophisticated integrated programme.

The same drawings arrive at the engineer's office, prompting the hassled engineer to call and ask for digital copies — the digital issue having been forgotten in the panic. By the time these have been successfully transferred, unzipped, translated into the right format and imported as base layers, it's ten days since the target issue date. Then, guess what? The engineer's drawings are late too, partly, as he bleats to an unsympathetic client, because the architect didn't get his act together early enough, partly because his plotter went down just at the wrong time and partly because ... I'm sure there's another excuse lurking in there somewhere. Then the cost report is late too, because the quantity surveyor was waiting for drawings, and can't get prices back from suppliers without decent information.

This is not an unusual procedure in construction. So why can't we manage to actually deliver on the dates that we agree? I suspect that the reason for failing on some occasions is due to a mismatch between those making the promises and those delivering them. The 'concept team' allocated to a project will often be the big cheeses, the articulate, graphically talented creative types who successfully convince the client that they've got the right firm for the job. Sure, this is probably the interesting part of the design process, drawing on inspiration, allowing imagination free rein, coming up with the 'big idea' — it's the 'fat felt pen' stage.

There's no doubt that some extraordinarily elegant solutions have emerged from broad-brush sketches on the back of napkins ('Millennium Bridge' for one). However, as the concept develops, this team is also required to formulate an outline plan for delivery, agree a budget, set some programme milestones. Is this the same team that will be churning out the drawings during the production information stage? Unlikely. And if not, then the handover between the concept and the delivery team needs to be a comprehensive transfer of information: not just about the materials, the appearance, the floor plan, but also the practicalities of when, where, how and — often forgotten — why.

If the delivery team is faced with a programme (or a budget for that matter) that is unrealistic, then better to shout early on and try to suggest mitigation measures than wait until the deadline is missed or the funding is spent. This sounds like common sense doesn't it, but you would be amazed at how often it doesn't happen though, leaving the client with a huge headache and the design team collectively shuffling its feet and shiftily flicking glances at each other as if to say, 'Who? Me, guv?'.

Contractors have the same problem in spades. The estimating team prices the job; the commercial director has said all the right things in the interview (whether true or not); and then the delivery team come on board. Faced with high expectations, any suggestion that budget or programme is under threat can result in bitter internal wrangles, or a conspicuous change in attitude, to the detriment of the job.

One supplier, who shall remain nameless, has demonstrated all the bad things that I'm talking about here. Asked to advise on a cladding solution for a modest new building, the masonry supplier suggested a highly innovative technique to the ambitious design team. The technique was sold as labour-saving and sustainable, and the design team duly incorporated the specialised masonry solution into the design. With a contractor on board, the supplier suddenly changes his tune. No, his system won't work in those conditions, can't be done before November and, by the way, is going to cost £30,000 more than originally quoted. Aghast, the design team try to contact the extremely plausible and personable sales director. Weeks of frustrated unanswered phone calls, illegible faxes and impenetrable messages leave both design team, client and contractor with a bad taste and the firm intention never to use the same supplier again. Ever. Anywhere.

There's no point in making unreasonable promises. You always get found out in the end — if not by your client, then by the rest of your own team. If you make a promise, stick to it, it's the only way to establish trust in working relationships.

Case Study 5

Making teams work

Why is it that the professionals in the property and construction industry are so much weaker at man-management than their counterparts in other industries? I'm talking about our ability to manage teams, to supervise people, to foster young talent. I can't speak for all the professions, and of course there are those exceptions that prove the rule, but an almost universal gripe from the rank-and-file within engineering consultancies is 'Well, their projects are really interesting and I enjoy the work, but their man-management skills are awful ...' — or words to that effect.

Are our management skills so poor? Is it because designers don't have the right sort of brains to be managers? Too creative? Too interested in the technical issues to worry about people issues? Surely it needn't be so.

At one level, there are the professional institutions. They provide a framework for training and acquiring a professional qualification. Each institution operates slightly differently, but all require that young aspirants to qualification (whether it be chartership or associateship or licentiate or whatever) are given some regular supervision by an active member of the Institute. This may be in the manner of a formal training agreement, with periodic submissions and checks, or it maybe a less formal, more infrequent review system.

The Institution of Civil Engineers runs perhaps the most exacting programme, with quarterly reports demanded in some detail from graduates. While this is can be a source of complaints ('I thought I'd given up writing essays when I left college ...'), there is no doubt that it helps to remind the young engineer of things learned, mistakes to be avoided, and focuses their attention perhaps on areas where they have not yet gained sufficient experience. Yes, it is time-consuming, particularly if idly put off until an end of year scramble, but an hour a week is not so hard to find, particularly if the working week is deliberately structured to accommodate this.

Other institutions operate an exam policy — the notoriously difficult seven-hour marathon run by the Institution of Structural Engineers for example — that requires revision and practice at past papers. Here too, the young engineers would benefit from some allocated time to study. There does seem to be a reluctance for some companies to acknowledge the importance of such a support structure, with appointments missed, revision periods squeezed into lunch-hours or pushed out to evenings. The argument seems to be that as responsible members of the workforce, these new professionals should be able to organise their own time to integrate the necessary study periods. Sometimes though, wouldn't it be a good idea to give them a shove in the right direction? Surely nurturing our young talent is worth sparing an hour a week?

Then there is in-house training. Many companies state smugly that they operate in-house training/mentoring/development programmes. It doesn't matter giving it a fancy name as long as it happens. How often does your boss sit down with you and ask, 'How's it going?', and actually expect a genuine answer rather than a platitude? How often do you get a chance to tell them how it *really* is, to share your worries and ambitions, without your boss looking anxiously at her (!) watch and making you feel that this is down there with emptying the septic tank on a list of things she'd rather be doing?

Exactly. Once a year, probably, at salary review time. Or possibly late at night at the pub after a works event, when you're sufficiently inebriated to vocalise all those irritants that you've been brooding about for months. It's not really the best method is it? And yet those same bosses complain that they can't find, or can't keep good staff.

Perhaps we need to take a step back, and examine which is more important — meeting that deadline, issuing that costplan, answering that fax, or making sure your most promising team member isn't about to join the competition because of something trivial ('I don't drink coffee and there isn't a tea urn'). Trivial things can be magnified into reasons for leaving if allowed to fester.

This is as important at the highest levels as it is at junior grades. Even directors need to have the opportunity to talk their problems through with a superior, to get some career guidance, to share resourcing difficulties, even — God forbid — to ask for help.

Yes, we work in a demanding, narrow-margin industry, where fees are discounted and overheads trimmed, where time spent on things other than fee-earning work is squeezed and criticised, where personnel in support and administration roles are resented, and where the end almost always justifies the means. But perhaps we should take some tips from those in the manufacturing sector, and realise that a contented workforce is a more efficient workforce.

Part of being content is understanding where you are, knowing where you're going, and how to get there. Or at the very least knowing who to ask for a map. Me? I'm off to draw up a route-plan for getting to the office party next week ...

Case Study 6

Talking technical

TLAs are getting OTT! Three letter acronyms are becoming so prevalent in our business that we have begun to speak a cryptic language of letters and codes, intelligible only to a few. Here is a sample real conversation, with annotations for the uninitiated.

'So, is everything AOK [alright and OK] with this PFI [private finance initiative] then?'

'We've just reached ITN [invitation to negotiate] stage, and we're wondering how best to leverage our USP [unique selling point] within the JV [joint venture] for maximum benefit. We hope to move on to BAFO [best and final offer] asap [as soon as possible].'

It seems that the private finance initiative, or public/private projects [PPPs] has introduced a completely new vocabulary as teams try to unravel the mysteries of the selection process. A cursory glance through any construction magazine (the NCE or BSJ anyone?) will reveal a whole host more TLAs with which we baffle ourselves daily, from the entirely recognisable VAT [value added tax], the straightforward R&M [repair and maintenance], the regularly eulogised KPIs [key performance indicators], through the titles of the CIC [Construction Industry Council] or CIB [Construction Industry Board] — now superseded by the ETB [Engineering and Technology Board] — to the downright unlikely RIW, which as far as I can ascertain is a tremendously successful weatherproofing product whose proper name is 'really is waterproof'.

In technical circles, TLAs are perhaps even more prevalent. Building services engineers are particularly prone to word contractions: so yes, there is a difference between an MCB [miniature circuit breaker] and an MCCB [moulded case circuit breaker]; there's the mundane sso [switched socket outlet] (which may be a tsso — a twin — or a ssso — a single switched socket outlet); the more obscure PIR [passive infra-red detector]; the impressive cable description PVC/SWA/PVC [poly-vinyl chloride/steel-wire armoured/poly-vinyl chloride] or XLPE/AWA/LSF [cross-linked poly-ethylene/aluminium wire armoured/low smoke + fume]; then the basic AHU [air-handling unit] or FCU [fan coil unit]; the list could go on and on. This is to add the already abbreviated world of contract documentation, such as JCT-80 — does anyone remember that this stands for Joint Contractors Tribunal 1980 version? which may require the application of LAD [Liquidated and Ascertained Damages]. Then there's the happy realm of quantity surveyor-measurement speak, where your BQs [bills of quantity] may be measured using SMM7 [Standard Method of Measurement 7], or perhaps CMS6?? Or even POMI [Principles of Measurement, International], and may be split into S&C [shell and core] and FF&E [fixtures, fittings and equipment].

Survey drawings are another rich source of TLAs. From the well-known AOD [above ordnance datum] which occasionally appears as ODN [ordnance datum newlyn] to the manhole mystifyingly labelled UTL, which I found out to my cost recently stands for unable to lift. Indeed, given abbreviated pipe descriptions, such as 225mm dia AWA SWS [225mm diameter Anglia Water Authority Storm Water Sewer], survey drawings, when issued without a key, become an impenetrable collection of apparently random letters and numbers scattered across the sheet like so many scrabble counters.

Units of measurement too will appear as abbreviations, such as kVA [kilo-volt amps] kPa [kilo-pascals] or mm/m [millimetres per metre]. Here, it's important to get the capitalisation right, given the range of meaning. For example, M indicates a thousand and m indicates a thousandth. Consistency in superscripting would be useful too: is cubic metres represented as m3 or m^3 or cum? Is kilogrammes per cubic metre shown as kg/m3, kg/m^3 or, bizarrely in my view, kgm-3?

So, is this something we need to really worry about? Perhaps not — assuming we are all talking the same TLA language. The difficulty arises where translation between languages is needed. Most of the abbreviations and acronyms noted above are common enough to be widely recognised, but every company develops their own shorthand.

Each company has its own SOPs [standard operating procedures] or QMS [quality management system]. So one architect I know uses DTM [design team meeting] and PTM [project team meeting] to distinguish between meetings where the client is present or not. It's helpful if you know this when you start working with these teams. Even within the same company, individuals will develop their own personal dialects, and understanding some electrical engineers can be as difficult as comprehending swift Cockney or broad Geordie.

I am not advocating the end to acronyms — documentation in this business is hefty enough without trying to expand all that shorthand — but simply suggesting a more considered approach to introducing new abbreviations. In trying to make our industry more co-operative not to mention more attractive to young people, elimination of unnecessary jargon must be a benefit. So come on, join the AAA [Against Acronym Abuse] campaign!

To email or
not to email?

Email is a wonderful invention isn't it? I mean, it saves all those tedious phone calls, it allows us to spend even more time welded to our computer screens and to giggle at pretty weak jokes. Email has become the means of communication of choice for a whole tranche of our industry.

- 'I'm a user, but I could give up at any time — honest.'
- 'I only use it for recreational purposes really.'
- 'It helps me get through the working day.'

Sounds to me as though our excuses demonstrate that emailing is becoming dangerously addictive. Maybe this is not something to worry over however. After all, it's not as though it's doing any physical damage to our bodies, unless you believe all those cranks that insist that radiation from computer screens is inexorably penetrating and affecting our brains. Or the folk that believe that staring at a VDU all day causes eye-strain. Or the whingers who complain that sitting too long in an office chair gives them backache. Or that typing has caused RSI (repetitive strain injury). So, our health is entirely unaffected by this growing indulgence, obviously.

What gets me really I suppose, is that so much of our email traffic is completely unnecessary. Does every single individual in a company really need to know that Lucy in Accounts in Basingstoke has lost an earring? Is it sensible to use an email system to ask for someone's whereabouts because there's a phone call on the first floor? Or to find out what's on the menu for lunch? The fact that the system is used for these things seems to imply several rather worrying trends:

1   people are expected to be sitting at their computers most of the time
2   people are expected to respond to that annoying beep that alerts them to the arrival of an email immediately, regardless of whatever task they're currently pursuing
3   people are happier using a keyboard than a telephone.

It's the last one that is perhaps the most worrying. I recently received an email from a colleague two floors down asking 'Are you in today?' Now why said colleague couldn't pick up the telephone to ask the question is completely beyond me. Surely it's simpler, quicker and much more likely to get an immediate response than an email. (At which point I have to admit that I emailed back 'No, this is a hologram'.)

Perhaps we are becoming increasingly reluctant to have anything to do with genuine human interaction, choosing to interface with a machine rather than, heaven forbid, actually talk to another person. As I've said before, there are so many things that can be conveyed via a conversation that are just not possible in electronic means — no matter how many smiley, winking or frowning character faces are used :-).

For dissemination of information, email systems are great: particularly if set up and used with a certain amount of rigour. Email conferences or notice-boards that list forthcoming events, or provide a forum for debate about current technical issues, or even allow requests for advice are genuinely useful tools, but it's no good if they're not known about, or misused. (Lucy could have limited her lost earring plea just to the Basingstoke office, for example.)

Conversely, too many refinements on offer can lead to 'option paralysis' as users are confronted with a whole tree of folders to access. The best thing about email must be its ability to allow dissemination of vast amounts of complex information to large numbers of people within a short space of time. Lengthy documents, that would otherwise take time (not to mention money) to photocopy, issue and post, can be sent down the line in the blink of an eye.

Drawings are a good example of information exchange between team members relatively quickly. (Although I've seen problems in this area too, as technicians struggle to download large files, convert formats or find cross-reference files). Providing companies share the same software — or at least have an IT department with the wherewithal to do translations — information exchange can be done entirely via email, with very little pain.

Which brings me to my last niggling worry — tracking and storage of information. In an industry that relies on accurate information in order to build things, can this wealth of data be managed as successfully in soft form, ie as a digital file, as it can in hard form, ie as a piece of paper? There's something rather nebulous about an email. If you're not copied, you don't know it's been sent, and there's no opportunity for finding out by spotting interesting bits of correspondence on colleagues' desks.

Chances are there's no filing system for email exchanges, that the company Quality Assurance Manual doesn't cover them, that older messages are auto-archived (as far as I can see, computer speak for 'chucked in the bin') after 100 days. The paperless office is all very well as an ideal, but I for one can't help but think that there's a certain solid reassurance to bulging manila folders, to letter-headed paper, to real ink signatures.

So yes, I'm a user, but I don't rely on it — no, really I don't.

Understanding property professionals

Understanding quantity surveyors

The construction industry has been around for a long time. Since early Wessex man successfully hauled 36 monoliths from a Welsh hillside and assembled them as a stone age alarm clock on flat patch north-east of Salisbury, construction has been a major employer and has made a significant impact on the way we live our lives.

And yet, in terms of the way we approach the task of building, curiously little has changed. The first architect was contributing to the Greek Empire and the term civil engineer dates from medieval times to distinguish the role from a military engineer. Institutions for architects, engineers, in all their various guises, and quantity surveyors have collectively been around for over a thousand years. These construction professionals have proved a hardy strain, well-adapted to the environment in which they work. Recently, a number of the established institutions have chosen to merge, in order to form a stronger representative body, with titles metamorphosing to suit: The Institution of Electrical Engineers (IEE) has merged with the Institution of Incorporated Engineers (IIE) to form the Institution of Engineering and Technology (IET). Contractors too have been around for a very long time, albeit surviving a number of mutations in the process: from the serfs that shifted stone for their Saxon lords to the masons that crafted the great cathedrals, the railwaymen that tamed the wildernesses right up to the specialist contractors, construction managers and management contractors of today. Builders have successfully evolved and proliferated to fill a changing need. So how should these strange species continue to survive? Is the population stable? What are the influences of habitat, climate?

A well-known, but nameless senior figure in the industry — one of our more enlightened clients — is fond of saying that he would welcome a virus that wiped out quantity surveyors as a species. The root of this vehement dislike appears to be the argument that they add nothing to the construction process — given a client to commission a building and a design team to deliver that building, why should a breed which has little to do with either be the one to say how much the building will cost? Now, don't get me wrong, some of my best friends are quantity surveyors, but I can sort of see his point. The days of 'surveying quantities' are fast disappearing, and the traditional quantity surveyor may be an endangered species.

The historic role

Once upon a time, when the design and construction process was a more considered and less stressful one, there was a clearly linear route from idea through concept design and detailed design to tender. At this point, the quantity surveyor stepped in to generate a set of documents that fully described the design in terms of 'stuff that needs to be bought': tonnes of concrete, lengths of steel, areas of brickwork, numbers of door handles, lengths of electrical cables or copper pipework, even numbers of bolts. This was a self-contained, and often quite separate duty — the quantity surveyor had no need to be familiar with the designs (although it was simpler if he was) as he received a complete design, architectural and engineering drawings and specifications. He closeted himself in his office, with drawings spread across the desk (or perhaps a digital measuring table if he was lucky) and measured each widget on the job. Carrying out this sort of analysis, even for a fairly modestly sized building, became a specialised task, with its own rules of measurement, organisational structure and terminology. A well-prepared Bill of Quantity required experience and accuracy and provided a key document in the procurement and management of the project. However, such complex documents require time to prepare, and clients seem to be unwilling to pay for the necessary time commitment from their quantity surveyor team.

Does this mean that we will soon be drying a tear over the grave marked 'Quantity Surveying, RIP 1792–2006'? I think the rumours of the profession's demise have been exaggerated: so 126-page Bills of Quantity may be a thing of the past, and the specialist measurement skills required are becoming increasingly redundant, but there is still a place for a cost expert on the team.

Finding a new place within the team

The difficulty that quantity surveyors face is the transitional period, as they move away from the passive historic role, towards a more active position within design teams. There is no doubt that an experienced quantity surveyor brings a valuable talent to the design process, able to advise instantly on relative costs of options under consideration and point out possible constructional difficulties. Yet this requires confidence, dynamic involvement and a willingness to be put on the spot as well as experience of construction techniques and technologies that are constantly changing. Unfortunately, these are qualities not often associated with the pin-stripe suited quantity surveyor stereotype — particularly with the advent of the litigious client, the reluctance to give on the spot advice, that may be challenged (or proved erroneous) becomes over-whelming. The complacent 'give me a drawing to measure' attitude is the cause of endless frustration among designers. In these days of fast-track projects and ever more demanding clients, there is simply not the time to draw every option to a level of detail where an estimator can count the bolts to build up a price.

So, can the quantity surveying profession evolve to match the changing construction climate? Many quantity surveyor practices have already evolved, reinventing themselves as cost consultants or construction cost advisers, or diversifying to become project managers, construction managers or total service providers offering 'professional management services to the industry' — not a word about cost in there at all. Then, is the profession in danger of evolving too far? If all surveyors miraculously metamorphose into general managers, who will remain to give informed advice on cost issues? One answer may be to rely on contractors for prices — after all, they are the ones who are really in touch with costs, so their advice must be more current and more reliable. This is laudable up to a point, but contractors may not be equipped to give cost advice at the very early stages of a project, nor may it be appropriate for them to be establishing cost plans on behalf of clients.

As the construction climate changes, more partnering arrangements are introduced and we see more collaborative relationships between clients, consultants, contractors and suppliers, the cost 'referee' role previously occupied by the quantity surveyor is being gradually sidelined. This, together with the erosion of traditional pricing techniques will almost certainly lead to fewer obvious openings for the quantity surveyor. However, the quantity surveyor still has a role to play, particularly in cost-planning at the early stages of projects, although a re-branding to suit a faster, more dynamic market might help in removing the lingering doubts about their value to the planet as a species.

Dealing with architects

Architects were once described as 'the pop stars of the construction industry'. They are the ones that front the projects, that have their name attached, that get feted on opening night, that graciously take the applause and the awards at the industry dinners. By contrast, other members of the construction team are relegated to the back of the stage, or even to the wings, fortunate to get a name-check at that awards ceremony, dismissed as 'the backing band'. Is this a fair view of the architecture profession? And if it is so, then how can other construction professionals work with these prima donnas to achieve the right solutions for their clients?

Signature architects

There is an intense debate within the architecture community about 'signature architects' and 'iconic architecture'. One view is that architecture that produces 'icons', highly recognisable architectural statements, is a distinctive and valuable contribution to our visual environment. The counter-argument is that the primary aim for any piece of architecture should be to deliver the appropriate functional requirements, and a secondary aim should be to fit appropriately within its context, and that grandiose statements are therefore anathema to truly great architecture. Whatever your view of this particular debate, the skyline of many of our cities is testament to a handful of architects that espouse the iconic rather than the discreet. The 'Gherkin' in the City of London (more properly called 30, St Mary Axe, or the Swiss Re building) — architect Norman Foster and Partners; the Millennium Dome (whatever you think of its political birth) — architect Richard Rogers; these two buildings are fine examples of that iconic tradition, instantly recognisable symbols of London.

Case Study 9

Why courtesy works

Manners is not a word used often these days — an occasional injunction to a naughty child perhaps — but as a social skill it seems all but forgotten. Now maybe it's just that I'm getting older and more intolerant, but in business, common courtesy and politeness seem to be vanishing over the distant horizon faster than Michael Schumacher's Ferrari.

Take this scenario: Client and engineer are having a face-to-face conversation about a tricky groundworks problem in the client's office. The client's telephone rings. In mid-conversation — in mid-sentence in fact — he swivels his plush leather chair around, turning his back on the engineer to answer the insistent ring, leaving the engineer to stand helplessly in the middle of the room. A few things strike me about this scene.

- No-one seems to get offered a seat anymore. When knocking on an office door, the response is more likely to be a grunt and a terse 'Whaddaya want now?' rather than 'Come in, have a seat, what can I do for you?'

- The client did not excuse himself to answer the telephone, or suggest that the conversation could be concluded later. Indeed, he could have ignored the telephone altogether, secure in the knowledge that a colleague, or the voicemail system would intercept the call. Still, it's difficult to ignore a ringing telephone — I suspect telecom companies have selected ringing tones that

275

subliminally whisper 'Urrr-gent, urrr-gent, urrr-gent ...', even when we know that it's more likely to be a double-glazing salesman than the Queen inviting us to dinner at Buckingham Palace.

- The client's attitude that his engineer's time is sufficiently valueless that several minutes shifting uncomfortably from foot to foot on his non-stain, logo-emblazoned carpet is time well-spent. Now I know it's the client's apparent prerogative to treat consultants like shop assistants, but I strongly suspect that the same client argues ferociously about the engineers' fees and wants to know exactly what he's been doing for all that time that he's charging.

Then there's other things — trivial perhaps, but noticeably absent from modern life.

- 'Please' has either been omitted from the vocabulary altogether, or has been replaced by an interrogative 'OK?' or 'yes?' at the end of a request. 'Look, the bloke's getting a bit desperate, make sure that drawing goes out on Thursday, OK?'

- It's amazing the difference an occasional 'thank-you' makes — no matter that it's part of someone's job description to, say, sort and fold 77 A1 drawings and put them into envelopes, a simple 'thanks' at the end of the rush at least acknowledges that it's been done by a person, not a machine.

- Is it physically impossible for some people to apologise? I do wonder sometimes. There are occasions when starting a sentence with 'I'm really sorry about this, but ...' can get you much further than even the most inventive excuse. (Even the most sympathetic of clients can get suspicious at the third grandma's funeral.)

- Punctuality — if a meeting is set for 10am, then it should start at 10am. I can't tell you the number of times I've seen the architect swan in at twenty past, with a casual 'we'd better start now'. Of course everyone else was just idling away the time, waiting for the most important person to grace them with their presence. Sure, there can be unavoidable delays, but isn't that one of the reasons mobile phones were invented?

- Holding doors open; 'after you'; a smile to acknowledge a service performed; offering to make the coffee (or to pour the tea — the manouverings at meetings to avoid 'being mother' are sometimes so intense that nobody gets tea at all).

These trivial examples of the lack of consideration for our professional colleagues is perhaps a symptom of a wider malaise. We seem so much wrapped up in ourselves, that any thoughtfulness towards others is simply alien. Thoughtlessness is becoming the norm.

Stop right there. What happens if we all stop thinking about what someone else wants? Stop trying to understand what our client is trying to achieve; stop thinking about how the architecture integrates with the engineering; stop wondering what this solution will cost; stop questioning how the contractor is actually going to build it — if that happens then we might as well all pack up our laptops and go home right now, because no-one will be able to successfully complete buildings anymore.

An extreme extrapolation you may think, perhaps, but one that merits consideration. At a time when we're trying to make the industry less confrontational and more co-operative, it's imperative that we do make an effort to think about what other parties in the process want. What drives them, their motivations, their needs? Maybe, just maybe, a little politeness will assist in defrosting some of those brittle, icy relationships and aid us in arriving at more collaborative working methods that allow us to build better, build quicker and build more profitably, and perhaps most important — to build together.

Index

For Product Safety Concerns and Information please contact our EU representative GPSR@taylorandfrancis.com Taylor & Francis Verlag GmbH, Kaufingerstraße 24, 80331 München, Germany

T - #0096 - 270225 - C0 - 216/138/16 - PB - 9780728205031 - Gloss Lamination